New Environmentally-Friendly and Sustainable Materials

New Environmentally-Friendly and Sustainable Materials

Editors

Alisson Mendes Rodrigues
Gelmires de Araujo Neves
Romualdo Rodrigues Menezes

MDPI • Basel • Beijing • Wuhan • Barcelona • Belgrade • Manchester • Tokyo • Cluj • Tianjin

Editors

Alisson Mendes Rodrigues
Federal University of Campina Grande
Brazil

Gelmires de Araujo Neves
Federal University of Campina Grande
Brazil

Romualdo Rodrigues Menezes
Federal University of Campina Grande
Brazil

Editorial Office
MDPI
St. Alban-Anlage 66
4052 Basel, Switzerland

This is a reprint of articles from the Special Issue published online in the open access journal *Sustainability* (ISSN 2071-1050) (available at: https://www.mdpi.com/journal/sustainability/special_issues/NewEnvironFriend_SustainableMaterial).

For citation purposes, cite each article independently as indicated on the article page online and as indicated below:

LastName, A.A.; LastName, B.B.; LastName, C.C. Article Title. *Journal Name* **Year**, *Volume Number*, Page Range.

ISBN 978-3-0365-3116-8 (Hbk)
ISBN 978-3-0365-3117-5 (PDF)

Contents

About the Editors

Alisson Mendes Rodrigues is currently a researcher at the Federal University of Campina at the Materials Technology Laboratory (Brazil). He is a multidisciplinary researcher with a chemistry degree from the State University of Maranhão (Brazil), an MEng in materials engineering from the Federal Institute of Education, Science and Technology of Maranhão (Brazil), and a D.Sc. in materials science and engineering from the Federal University of São Carlos (Brazil). He completed his doctoral internship at the University of California (EUA), worked as a researcher at the University of Rostock (Germany), and was a researcher at the São Carlos School of Engineering, which is part of the University of São Paulo (Brazil). Among his scientific interests, the following stand out: glass science covering fundamental and applied research on isothermal and non-isothermal crystallization kinetics in stoichiometric and non-stoichiometric glasses, glass technology in the development and characterization of bioactive glasses and glass ceramics, the manufacture of ceramic coatings by physical vapor deposition (PVD) techniques such as high-power impulse magnetron sputtering (HiPIMS), and the development of eco-friendly and sustainable materials from urban and mining waste.

Gelmires de Araujo Neves is a full professor at the Federal University of Campina Grande (UFCG), located in Brazil. He holds a BEng degree in mining engineering from the Federal University of Paraíba (Brazil), an MEng degree in chemical engineering from the Federal University of Paraíba, and Dr-Ing in process engineering from the Federal University of Campina Grande. He has experience in the areas of materials and metallurgical engineering, with emphasis on ceramic materials and material recycling, working mainly on the following topics: waste recycling, construction materials, ceramic material processing, organophilic clays for use in drilling fluid, iron ore pelletizing, adsorption of dyes and nanometric fillers, and characterization of ceramic raw materials. He has also coordinated several research and development projects funded by various funding agencies and companies, has published 211 articles in indexed journals, and has supervised hundreds of students (graduate, master, and doctoral level). He was the head of the Department of Materials Engineering, a member of the Full Board of the UFCG, the coordinator of the Postgraduate Program in Materials Science and Engineering, and a member of the Postgraduate Chamber of the UFCG.

Romualdo Rodrigues Menezes is an associated professor and the head materials science and engineering graduate program at the Department of Materials Engineering, Federal University of Campina Grande (Brazil). Before, he was an assistant professor of materials engineering at the Federal University of Paraiba (Brazil). He holds a BEng degree in materials engineering from the Federal University of Paraiba, an MEng degree in chemical engineering from the Federal University of Campina Grande, and D.Sc. in materials science and engineering from the Federal University of São Carlos (Brazil). The research activities of Prof. Menezes are in the broad areas of ceramic processing, nanotechnology, biomaterials, and waste recycling. He is the author or co-author of more than 150 scientific papers. He is a member of the Brazilian Ceramic Association and the Latin American Society of Artificial Organs, Biomaterials and Tissue Engineering.

Preface to "New Environmentally-Friendly and Sustainable Materials"

Nowadays, one of the key challenges facing contemporary society is the establishment a sustainable lifestyle that is in harmony with the environment and that does not compromise economic growth and technological development. Furthermore, the increase in the world population has boosted per capita consumption and, consequently, the generation of different types of waste. When irregularly discarded, waste can contaminate water, air, and soil and can cause damage to fauna and flora. Currently, billions of tons of waste are generated each year on a global scale, causing profound environmental impacts. Population growth tends to aggravate this scenario, and according to the United Nations, it is estimated that the world population will reach 9 billion in 2050, and if there are not changes in our current consumption patterns, then the expectation is that the amount of generation of waste that is generated will increase by several billion tons.

Because of this scenario, scientists and researchers have made a continuous effort to develop new materials from different types of residues. The goal is for these materials to perform is a way that is equal to or that is superior to materials that can already be found on the market, but with the seal of being more environmentally friendly and sustainable. This book is a print edition of the Special Issue entitled "New Environmentally Friendly and Sustainable Materials" that was published in Sustainability (ISSN 2071-1050). The aim of this Special Issue was to highlight and share recent scientific discoveries in the development of new environmentally friendly and sustainable materials created from different types of waste, emphasizing the compromise between chemical resistance, morphology, and its properties. We hope that readers will enjoy this collection and that it will serve as an inspiration for the emergence of new trends as well as scientific and technological advances.

Alisson Mendes Rodrigues, Gelmires de Araujo Neves, Romualdo Rodrigues Menezes
Editors

sustainability

MDPI

Article

Durability of Sustainable Ceramics Produced by Alkaline Activation of Clay Brick Residue

Fabiana Pereira da Costa [1], Izabelle Marie Trindade Bezerra [1], Jucielle Veras Fernandes [1], Alisson Mendes Rodrigues [1,2,*], Romualdo Rodrigues Menezes [1,2] and Gelmires de Araújo Neves [1,2]

1 Graduate Program in Materials Science and Engineering (PPG-CEMat), Federal University of Campina Grande, Campina Grande 58429-900, Brazil; fabiana.costa@estudante.ufcg.edu.br (F.P.d.C.); izabelle_marie@yahoo.com.br (I.M.T.B.); jucielle_fernandes@hotmail.com (J.V.F.); romualdo.menezes@ufcg.edu.br (R.R.M.); gelmires.neves@ufcg.edu.br (G.d.A.N.)

2 Laboratory of Materials Technology (LTM), Department of Materials Engineering, Federal University of Campina Grande (UFCG), Campina Grande 58429-900, Brazil

* Correspondence: alisson.mendes@professor.ufcg.edu.br

Abstract: Alkali-activated materials (AAMs) were produced using residues from the red ceramic industry as a precursor, and sodium hydroxide (NaOH), potassium hydroxide (KOH), and sodium silicate (Na_2SiO_3) as alkaline activators. The effect of activators and curing conditions on physical-mechanical properties and durability were evaluated. The processing parameters (amount of water and consistency index) and the activation conditions (the activator contents and curing temperature) were defined based on an experimental design getting the flexural rupture module as the response. The durability behavior was evaluated by natural aging, accelerated aging (simulated rain test), exposure to the marine environment (salt fog), and acidic environments (HCl and H_2SO_4). The results showed that the NaOH- and KOH-activated samples exhibited inferior mechanical behavior than those activated with Na_2SiO_3. In the durability studies, due to leaching, there was a decrease in mechanical strength when samples are subjected to aggressive exposure conditions. However, the strength values are still higher than the minimum indicated for traditional ceramic applications.

Keywords: ceramic residue; alkaline activation; experimental design; durability tests

Citation: Costa, F.P.d.; Bezerra, I.M.T.; Fernandes, J.V.; Rodrigues, A.M.; Menezes, R.R.; Neves, G.d.A. Durability of Sustainable Ceramics Produced by Alkaline Activation of Clay Brick Residue. *Sustainability* **2021**, *13*, 10931. https://doi.org/10.3390/su131910931

Academic Editor: Grigorios L. Kyriakopoulos

Received: 9 August 2021
Accepted: 6 September 2021
Published: 1 October 2021

Publisher's Note: MDPI stays neutral with regard to jurisdictional claims in published maps and institutional affiliations.

1. Introduction

The red ceramic industry is an important economic segment of the global production chain and generates several jobs. However, vast amounts of residues generate from such industry [1]. Generally, the red ceramic residues generated before the sintering step are reintroduced to the production process, but the residues generated after the sintering step are often discarded in inappropriate places causing damage to the environment. In the past decades, the environmental awareness of society has grown, and several studies have focused on developing new materials from the reuse of industrial waste to contribute to sustainable development [2–6].

The residues of the red ceramic industry have several potentials for reuse; among them stand out the production of traditional ceramics or cementitious materials [7–10]. Additionally, such residues have the potential to be used as precursors of alkali-activated materials because they are rich in silica and alumina and compounds by a large number of amorphous phases [11]. In recent decades, alkali-activated materials (AAMs) have attracted worldwide attention for their low CO_2 emissions and excellent mechanical and durability properties. These materials are formed from reactive aluminosilicate raw materials that, in an alkaline medium, undergo a series of chemical reactions giving rise to alkaline aluminosilicate gels with cementitious properties [12].

Due to environmental and technological advantages, AAMs have been widely used in the civil construction sector as an alternative to the use of Portland cement [13], as complementary materials to avoid the alkali-aggregate reaction [14] or in the production of ceramic parts without the sintering step [15]. Among the main properties, we can mention high mechanical resistance, more excellent chemical stability, slight curing retraction, greater thermal resistance, more outstanding durability in aggressive environments, and lower CO_2 emissions for its production than Portland cement, which is one of the most CO_2 emitting in the world [16,17].

Sodium silicate or alkali metal hydroxide (Na or K) are frequently used activators to produce alkali-activated materials [18]. The most commonly used precursors are materials that have good pozzolanic activity when mixed with Portland cement, such as fly ash [19], blast furnace slag [20], and calcined clays [21]. Additionally, several by-products and industrial wastes have been used as precursor material [22], such as red mud from the bauxite beneficiation process [23], rice husk ashes [24], glass waste [25,26], wastes from ceramic production [18,27–29], among others.

Rego et al. [30] studied various industrial residues, including ceramic residues as a precursor material in synthesizing geopolymer cementitious matrices using hydroxide and silicate activators. The authors observed that the geopolymers produced could be used as adhesives on metal plates subjected to temperatures up to 400 °C, showing better efficiency when compared to epoxy-based resins (adhesion resistance up to 200 °C). Azevedo et al. [31] characterized clay brick residues and evaluated their use as an alternative precursor for producing ceramic tiles through geopolymer reactions. The results showed that the ceramic residues showed high pozzolanic reactivity and are rich in silica and alumina, which are fundamental for synthesizing AAMs. Regarding technological analysis, the authors found that the water content and the curing temperature are decisive for improving the mechanical and durability properties of the AAMs.

Although the use of red ceramics residues as a precursor in the production of AAMs has already been reported in the literature, these are still little explored, especially concerning the aspects of durability in aggressive environments, such as seawater and acid exposure. In this sense, this work aimed to produce ceramic pieces through the process of alkaline activation using residues from the red ceramic industry and to evaluate the durability of the pieces formed from tests of natural aging, accelerated aging, and exposure to the marine and acidic environments (hydrochloric acid (HCl) and sulfuric acid (H_2SO_4)). For this, three different alkaline agents (NaOH, KOH, and Na_2SiO_3) were used as activators. The effect of activators and curing conditions on physical-mechanical properties and durability were evaluated.

2. Materials and Methods

2.1. Materials

Red ceramic residues (RCR) used in this work came from a commercial brick plant located in the state of Paraíba, Brazil. The residues were ground and sieved for different sizes 0.42 mm, 0.18 mm, and 0.074 mm. The alkaline solutions were obtained from sodium hydroxide (PA 97%, Synth, Diadema-SP, Brazil), potassium hydroxide (PA 85%, Synth, Diadema-SP, Brazil), and sodium silicate (Na_2O and SiO_2 = 20–30%, H_2O = 40–60%, VETEC, Duque de Caxias-RJ, Brazil) with basic pH (11–13). Ammonium polyacrylate was used as a dispersant (0.5 wt.% Lioperses 511, Miracema-Nuodex Indústria Química Ltd.a, Campinas—SP, Brazil), and ammonium hydroxide (30–32% PA, VETEC, Duque de Caxias-RJ, Brazil) was used to adjust the pH.

2.2. Formulation of the Granulometric Compositions

The samples were obtained from a combination of three granulometry: #40 (0.18 mm < particles size $\leq$ 0.42 mm), #80 (0.074 mm < particles size $\leq$ 0.18 mm), #200 (particles size $\leq$ 0.074 mm). Initially, the amount of water needed to obtain a consistency index of 260 $\pm$ 5 mm [32] was determined using the centroid simplex-type mixture design methodology, augmented with internal points [33]. Two replicates were performed for each point (three measurements). Table 1 presents the formulations corresponding to the points defined in the experimental design.

Table 1. Formulations of the granulometric compositions, obtained from the experimental design.

Formulations	Fractions (%)		
	#40	#80	#200
C1	100	0	0
C2	0	100	0
C3	0	0	100
C4	50	50	0
C5	50	0	50
C6	0	50	50
C7	33.33	33.33	33.33
C8	68	16	16
C9	16	68	16
C10	16	16	68

Since the minimum amount of water to reach the consistency index of 260 ± 5 mm was determined, another experimental design was accomplished to determine the consistency index. Such analysis was performed using the same data shown in Table 1. Dispersing and pH control additives were used to obtain the formulations. The significance and prediction of the models were evaluated based on statistical analyses. Due to better workability, composition with the highest consistency index was chosen for alkaline activation studies [34].

2.3. Characterization of the Red Ceramic Residue and Activated Samples

The chemical composition was defined by X-ray fluorescence (Shimadzu, EDX 720). The mineralogical phases were identified using the X-ray diffraction technique (Shimadzu, XRD6000), with Cu-Kα 2θ radiation in the range of angles of 5°–70°, 0.02° step and JCPDS database.

2.4. Alkaline Activation, Curing Times, and Preparation of the Samples

An experimental design of the 2 × 2 factorial type, augmented with star points and a central point was carried out for each activator (NaOH, KOH, Na_2SiO_3) considering the variables of cure temperature and activated concentration. Table 2 presents the experimental design used. The temperatures used in such analysis were based on the works of Rovnaník [35], Medri et al. [36], Bakri et al. [37], and Heah et al. [38]. For calculations of the alkaline activator's concentration, it was considered that all silica is reactive and originated from the kaolinite of the precursor material. The calculations were determined from the chemical composition of the red ceramic residue.

Table 2. Curing temperatures and activator alkaline activator's concentration obtained from experimental design.

Conditions	Temperature (°C)	Concentration (g/100 g)		
		NaOH	KOH	Na_2SO_3
1	21.8	9.0	14.0	30.0
2	30	6.0	9.0	20.0
3	30	12.0	19.0	40.0
4	50	4.77	6.93	16.0
5	50	9.0	14.0	30.0
6	50	13.23	21.07	44.1
7	70	6.0	9.0	20.0
8	70	12.0	19.0	40.0
9	78.2	9.0	14.0	30.0

The mixtures of the red ceramic residue (precursor) with the solutions of each alkaline activator (NaOH, KOH, and Na_2SiO_3) were homogenized in a mechanical mixer for 5 min. After that, the homogenous mixture was poured into cylindrical molds with 2.5 cm × 5 cm (diameter x height) and curing was carried out under temperatures defined in the experimental design (Table 2). The experimental conditions of curing used were 3 days (3d), 5 days (5d), and pre-curing for 4 days in the environment (E), followed by curing under the conditions of the experimental design for 1 day ($4d^E$ + 1d). The cures at 3d and $4d^E$ + 1d for Na_2SiO_3 samples were not performed due to difficulties in demolding the samples.

2.5. Durability of Samples after Alkaline Activation

Four tests were accomplished to investigate the durability behavior: natural aging, accelerated aging, exposure to the marine environment (salt fog), and acidic environments (HCl and H_2SO_4). For the natural aging test, the samples were kept in a closed place and protected from climatic variations in which the temperature variation (~22 °C–33 °C) and the air humidity (~75%) were kept constant. The samples were evaluated at 28, 60, and 90 days. The accelerated aging test was carried out through a simulated rain test, with 2 h wetting cycles and drying for 48 h in an oven (temperature 35 °C). Twenty-four cycles were performed.

Regarding the marine environment test, the samples were exposed in a coastal area, 100 m from the coastline for 28, 60, and 90 days to salt spray. For the acid resistance test, the samples were immersed in solutions (5% concentration) of hydrochloric acid (HCl) and sulfuric acid (H_2SO_4) for 28, 60, and 90 days. Then the samples were dried in an oven (35 °C) for 48 h. After the curing and durability tests, three points flexural strength (FS) tests were performed. Such experiments were accomplished in Shimadzu, AG-X-50 KN, with a speed of 0.5 mm.min^{-1}. The experimental values showed in this work correspond to the average value of ten samples.

3. Results and Discussion

3.1. Amount of Water and Consistency Index

Figure 1a,b shows 3D response surface plots, obtained from the cubic regression model, for amount of water and consistency index versus RCR formulations with different granulometric fractions: #40 (0.18 mm < particles size ≤ 0.42 mm), #80 (0.074 mm < particles size ≤ 0.18 mm), #200 (particles size ≤ 0.074 mm). Statistical treatment of the experimental data was performed to verify the significance and prediction of the experimental design. The cubic regression model was the most appropriate (95% confidence) to represent the amount of minimum water and the consistency index. Total of 320 mL was the minimum amount of water to obtain a consistency index of 260 ± 5 mm. Such value was observed in the region highlighted in Figure 1a (red hyperbole). This was because the RCR formulations contained in these regions probably favored better particle packing. Then, for the consistency index evaluation, the amount of water was kept constant at 320 mL. The part highlighted in Figure 1b (blue hyperbole) presents values of consistency index ranging from 240 to 280 mm, close to the value of 260 ± 5 mm recommended by the ASTM C1437 standard [32]. RCR formulations contained in this region have higher amounts of #80 and #200 fractions.

According to the results of the experimental design carried out for the amount of water and the consistency index, the Z composition (40% of #40, 2.5% of #80, and 57.5 of #200) was chosen to evaluate under which cure temperatures and alkaline activators concentrations obtained the best flexural strength values.

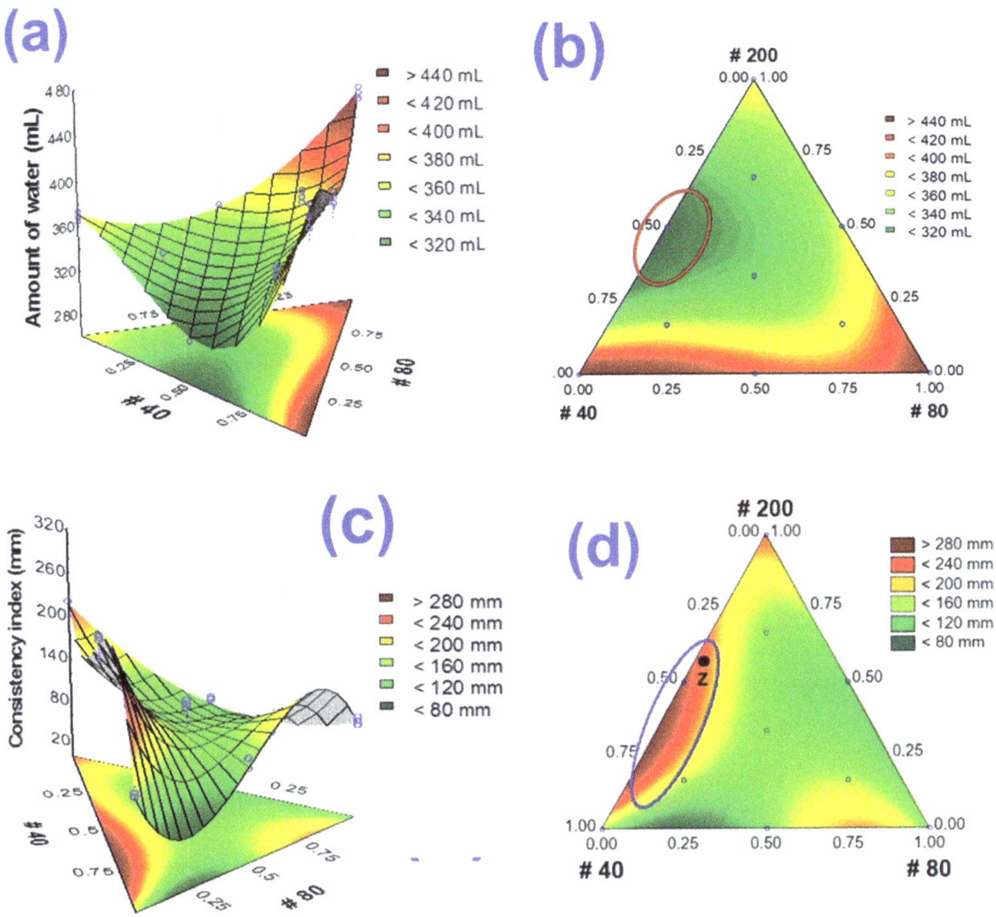

Figure 1. 3D response surface plots and their respective projections onto the composition triangle for the amount of water (**a**,**b**) and consistency index (**c**,**d**).

3.2. Curing Temperature and Alkaline Activators Concentration Influence on Flexural Strength

Figure 2a–c shows the 3D response surface plots for the flexural strength versus concentration of alkaline activators (NaOH, KOH, and Na_2SiO_3) and curing temperatures. The curing conditions 3d, 5d, and $4d^E$ + 1d were evaluated for NaOH, KOH while for the Na_2SiO_3, only the 5d was performed due to the difficulties in demolding the samples the other experimental conditions. For all conditions, the 3D response surface plot was obtained from the quadratic model (95% confidence). The coefficients of determination (R^2) obtained for curing conditions 3d, 5d, and $4d^E$ + 1d were 73.71%, 90.23%, and 92.20% for the samples activated with NaOH and 62.54%, 90.31%, and 70.33% for the activation with KOH. Finally, for the Na_2SiO_3 activator, R^2 was 86.92% (5d cure).

The highest FS values were observed at high cure temperatures and NaOH concentrations (Figure 2a). At intermediate temperatures (~50 °C), the NaOH activator did not significantly influence FS values in 3d and 5d cures. However, for the $4d^E$ + 1d cures, the FS values increase with NaOH concentration. In general, in the NaOH-activated samples, FS values ranged from 0 to 2 MPa for 3d, 0 to 3 MPa for 5d, and 0 to 2.5 MPa for $4d^E$ + 1d.

The KOH-activated samples (Figure 2b) show similar behavior as those activated with NaOH (Figure 2a); i.e., high FS values occur at higher cure temperatures and alkaline activator concentrations. However, at 3d cure condition, high FS values were obtained at low KOH concentrations and high cure temperatures. For the Na_2SiO_3-activated samples (Figure 2c), the highest FS were obtained for intermediate values of cure temperatures (~50 °C) and activator concentration (~30 g/100 g). Under these experimental conditions, sodium silicate dissolves silica, which reacts with Al^{3+} ions, forming aluminosilicate oligomers that unite, promoting the system gelation. It is known that the increase in gel volume favors the increase in mechanical strength [39]. Still for Na_2SiO_3, low FS values were observed for higher Na_2SiO_3 concentrations. It is believed that high Na_2SiO_3 concentrations negatively affect the synthesis kinetics, delaying the Si dissolution. As a consequence, the hardening of the system will occur when the gel only forms small agglomerates, which compromise the material's mechanical strength [40].

(a) **NaOH**

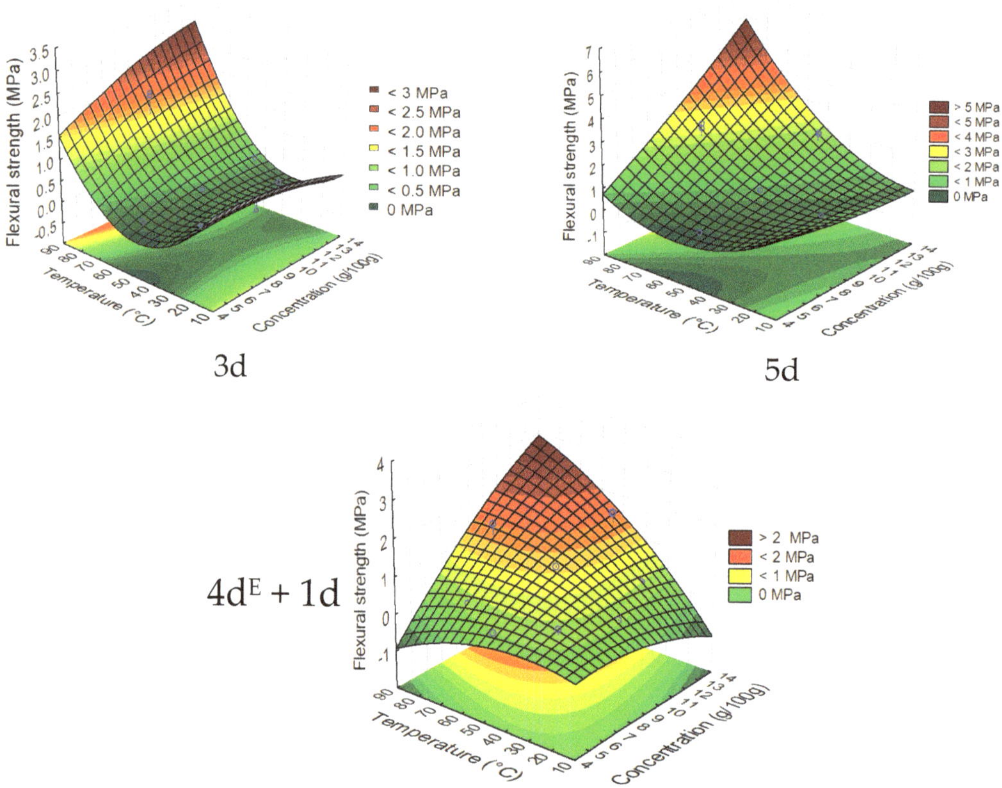

Figure 2. *Cont.*

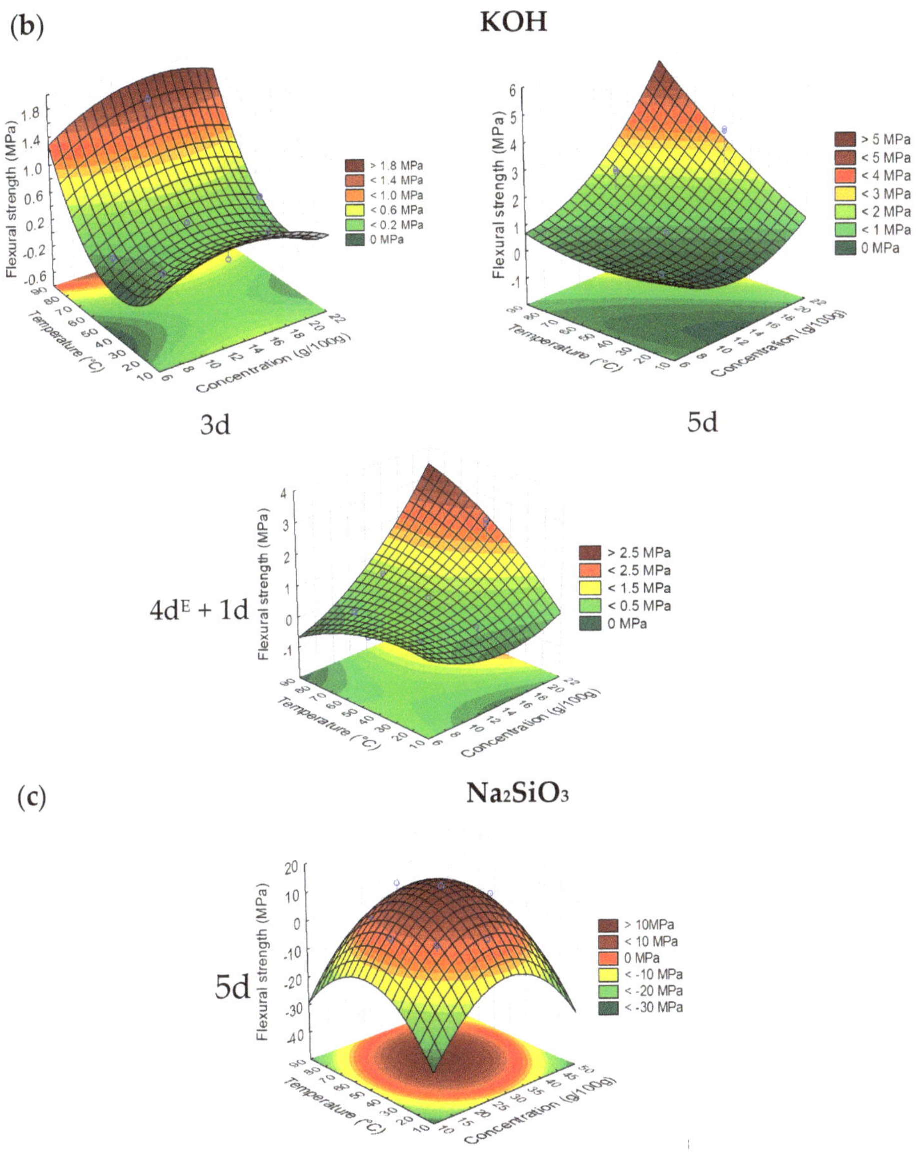

Figure 2. 3D response surface plots obtained from the experimental design of the activators and their cure times. NaOH (**a**), KOH (**b**) e Na_2SiO_3 (**c**).

According to the 3D response surface plots shown in Figure 2a–c, RNa, RK, and RNaS samples were chosen for the next experiments. The experimental conditions of these samples are summarized in Table 3. The 5d cure was chosen due to their better reproducibility for all studied activators (NaOH, KOH, and Na_2SiO_3); since the Na_2SiO_3-activated samples subjected to 3d and $4d^E$ + 1d cures presented difficulties for demolding. In general, the NaOH- and KOH-activated samples exhibited lower FS values than those activated with Na_2SiO_3. Such behavior may be related to large Si ions contents available in the early stages of geopolymer reactions, which accelerate their kinetics. Indeed, some research [41,42] found that activation solution contained Na_2SiO_3 which favored the geopolymer reactions, hence improving the mechanical properties.

Table 3. Curing conditions of the RNa, RK, and RNaS samples and their predicted flexural strength values.

Samples	Processing Conditions			Cure Type		
	Activator	[1] C, g/100 g	T, °C	3d	5d	$4d^E$ + 1d
					FS Value [2]	
RNa	NaOH	11.08	75	1.03–1.88	2.33–3.18	1.64–2.16
RK	KOH	17.13	75	0.66–1.52	1.44–3.22	1.10–1.96
RNaS	Na_2SiO_3	33.04	55	-	10.41–14.96	-

[1] Concentration; [2] predictive FS values acquired from Figure 2a–c.

3.3. Characterization of the Compositions

Table 4 and Figure 3a,b show the chemical composition and mineralogical phases analysis of red ceramic residue (RCR) and the alkaline-activated samples (RNa, RK, and RNaS). As discussed earlier, all samples were cured under 5d conditions. The SiO_2 and Al_2O_3 were the main oxides detected (44–51 wt% of SiO_2 and 20–26 wt% of Al_2O_3). Higher SiO_2 and Al_2O_3 contents (SiO_2 + Al_2O_3 = 77.3 wt%) in red ceramic residues favored the development of the materials with good mechanical resistance through the process of geopolimerization [43,44]. As expected, the Na_2O and K_2O contents varied according to the alkaline activator type used. In general, the mineralogical phases of all activated samples (Figure 3b) were the same as those identified in the ceramic residues (Figure 3a), mica ($KMg_3(Si_3Al)O_{10}(OH)_2$, JCPDS 83-1808), magnesium silica-aluminate, calcium and sodium (JCPDS 23-1405), feldspar (KSi_3AlO_3, JCPDS 84-0710), quartz (SiO_2, JCPDS 46-1045), and iron oxide (Fe_2O_3, JCPDS 04-0755).

Table 4. The chemical composition of the red ceramic residue (RCR) and of activated samples (RNa, RK, and RNaS).

Samples	Oxides (wt%)								
	SiO_2	Al_2O_3	Fe_2O_3	Na_2O	K_2O	MgO	CaO	TiO_2	Other
RCR	51.2	26.1	9.7	1.0	3.8	3.2	1.7	1.0	1.0
RNa	46.4	23.2	10.0	11.5	3.7	1.8	1.7	1.0	0.7
RK	44.3	21.9	10.3	-	18.1	2.1	1.8	1.2	0.3
RNaS	48.5	24.7	10.9	5.6	3.9	2.5	1.9	1.1	0.9

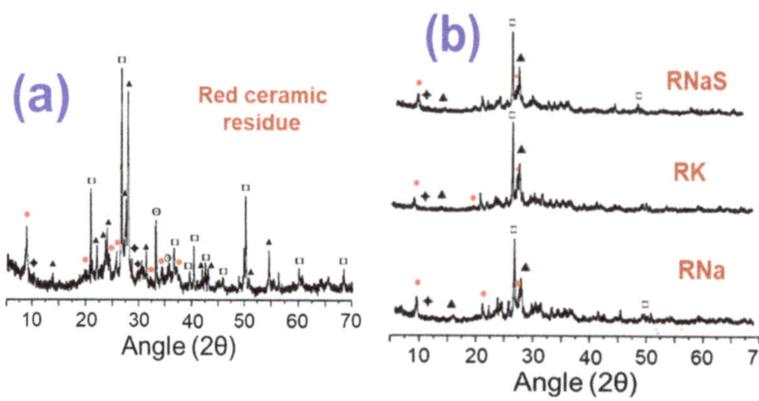

Figure 3. X-ray diffraction pattern measured from of the red ceramic residues (**a**) and the activated samples (RNa, RK, and RNaS) (**b**) (• mica, ✦ magnesium silica-aluminate, ▲ feldspar, □ quartz, ⊙ iron oxide).

3.4. Durability of Activated Samples

Figure 4 shows the FS results of RNa, RK, and RNaS samples after natural aging (Figure 4a), accelerated aging by simulated rain (Figure 4b), exposure to salt spray (Figure 4c), and exposure to hydrochloric and sulfuric acids (Figure 4d,e). For the Na_2SiO_3-activated samples (Figure 4a), was observed the trend of FS values increase around 60 days of exposure and then decrease up to 90 days. For NaOH-activated samples, it was observed that the FS values increased with the exposure time. In the case of KOH, due to the samples being practically deteriorated (Figure 5a,b), it was impossible to measure the FS values at 90 days. Such degradation behavior is related to the sub-efflorescence phenomenon (Figure 5b). In this phenomenon, under adequate humidity, unreacted K^+ ions leached toward the sample surface and depositing themselves in the salt form of salt.

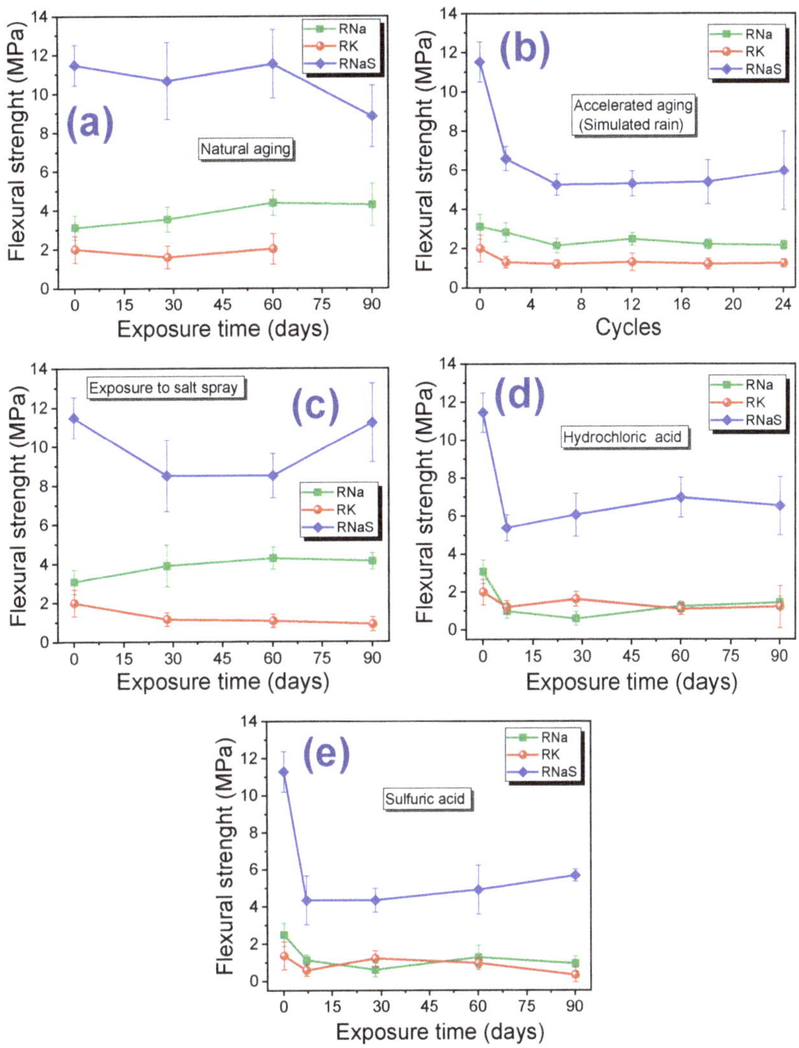

Figure 4. Flexural strength of the NaOH-, KOH-, and Na_2SiO_3-activated samples subjected to natural aging (**a**), accelerated aging by simulated rain (**b**), exposure to salt spray (**c**), to hydrochloric acid (**d**), and sulfuric acid (**e**).

Figure 5. Evolution of the deterioration process of KOH-activated samples after 60 days of natural aging (**a**), and efflorescence phenomenon exhibited by the RK sample (**b**).

For samples subjected to accelerated aging by simulated rain (Figure 4b), from 0° to 2° cycles, a significant decrease in FS values was observed only in the Na_2SiO_3-activated sample. Such behavior must be related to the leaching of soluble silica [45]. From the 2° to 6° cycles, the FS values decreased for all samples, regardless of the activator used, obtaining a subtle increase in the 12° cycle. This behavior may be related to the initial leaching of soluble particles present in the material structure. However, such behavior stabilizes with the continuation of the wetting and drying cycles, either by decreasing the ions leaching rate or increasing geopolymerization over cycles. From the 12° cycle onwards, NaOH and KOH-activated samples showed similar behaviors, indicating a slight decrease in resistance for the 18° and 24° cycles.

Figure 4c shows the FS values of alkali-activated samples that were exposed to salt spray for 28, 60, and 90 days. The mechanical resistance measured from NaOH and KOH samples did not undergo significant changes during the experiment. Unlike what was observed in natural aging, the sub-efflorescence phenomenon was not observed in the KOH sample; thus, FS values were measured at 90 days. For Na_2SiO_3-activated sample, from 60 to 90 days, there is a considerable increase in FS values. Probably this behavior is related to the continuation of the geopolymerization reactions or due to variations in the environment humidity, which favored the dissolution reactions and condensation. For the samples exposed (immersed) in HCl and H_2SO_4 solutions (Figure 4d,e), it was noted that the NaOH- and KOH-activated samples presented similar durability behavior, and there are no significant variations in FS values for these samples. For Na_2SiO_3-activated samples, there is a considerable decrease in resistance in the period from 0 to 7 days. After 7 days, the resistance tends to stabilize with the time of exposure.

Figure 6 compares the FS values measured from RNa, RK, and RNaS samples with natural aging with different geopolymers found in the literature. The following materials were collected from the literature: commercial metakaolin (GM) activated with sodium silicate/sodium hydroxide solution [35], ceramic residue (GPM 0.6)-activated alkaline activated with sodium silicate/sodium hydroxide solution [46], chamotte residues molded and pressed (CM and CP, respectively) alkali activated with potassium hydroxide/sodium silicate solution [31], Metakaolin HP Ultra molded and pressed (MM and MP, respectively) alkali activated with hydroxide/sodium silicate solution [31], fly ash with a proportion of 15% granulated blast furnace slag (AA3) alkali activated with sodium silicate solution [47], blast furnace slag and fly ash (50FA1.00) alkali active with Na_2SiO_3/NaOH solution [48], geopolymeric concrete prepared using a Class F fly ash (A40 S00), and alkali activated with 114.3 kg/m^3 Na_2SiO_3 and 45.7 kg/m^3 NaOH solution and 100 kg/m^3 of Na_2SiO_3 and 40 NaOH kg/m^3 solution (A35 S00) [49], fly ash-based mixtures (AF-AS) and ground granulated blast furnace slag (GGBS) alkali activated with sodium hydroxide/sodium silicate solution [50], low-calcium fly ash (GPC) alkali activated with (103 kg/m^3) sodium hydroxide/(41 kg/m^3) sodium silicate solution [51].

The highest FS values were found for Metakaolin HP Ultra pressed (MP: 30.44 MPa) and molded (MM: 12.50 MPa) residues [31]. In this case, the proper cure temperature was also investigated, and it was observed that the aluminosilicate dissolution reactions reached a maximum at 60 °C. RNaS sample, in this work, presented values around 11.5 MPa, in cures of 0 and 60 days, which were superior to other materials. RK and RNa samples in this work showed FS values ranging between 2 and 4.2 MPa, being lower than other strengths shown in Figure 6, except for the AA3 and GGBS samples. Sample AA3 was activated with 8% sodium silicate content, and the authors observed that for lower levels, the FS values decreased. According to the authors, the FS value for the AA3 sample (2.1 MPa) was satisfactory. GGBS sample showed the lowest FS result compared to other materials analyzed. This behavior occurs due to the fragile structure matrix produced from GGBS.

In general, it is possible to observe the increase of FS with the curing time; this behavior is expected because, for them to occur as geopolymerization and compaction reactions of the specimens in a satisfactory way, an adequate curing time is necessary [31]. However, for the samples in this work, the highest resistance was observed in 60 days of exposure for RNa and RNaS and 28 days of exposure to RK. Probably the geopolymerization reactions were consolidated at these ages.

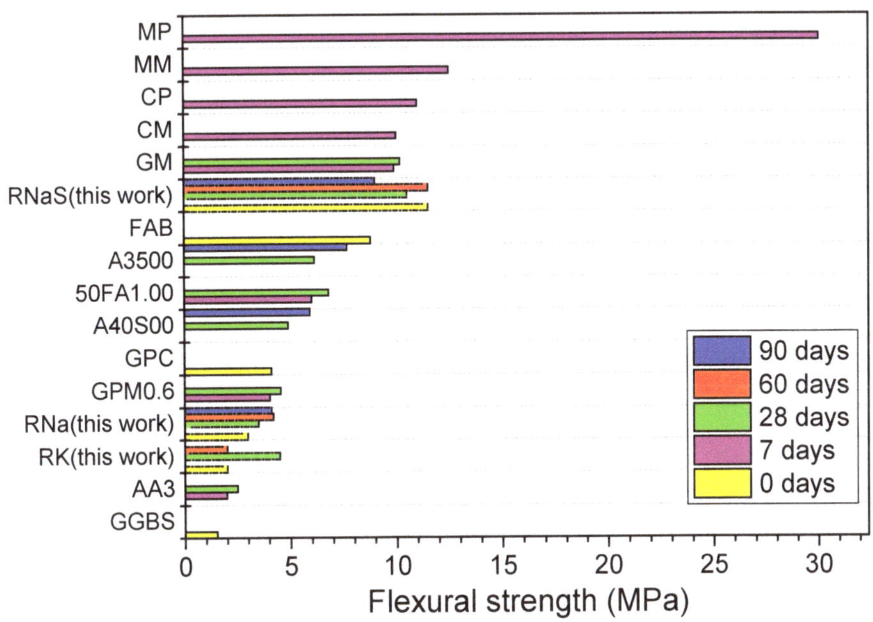

Figure 6. Comparison of the flexural strength of the red ceramic residues activated with NaOH (RNa), KOH (RK), and Na_2SiO_3 (RNaS) evaluated in this study with different geopolymers found in the literature.

4. Conclusions

The red ceramic residues showed potential to be used in the alkaline activation technique to obtain geopolymers. Curing conditions (temperature and time) and types and concentrations of the alkaline activator influenced the samples' mechanical strength and durability. The samples activated with NaOH and KOH exhibited less mechanical behavior than the samples activated with Na_2SiO_3. For NaOH and KOH activators, the highest flexural strength values were obtained with the combination of elevated temperatures and higher concentrations of the activator. In the case of the Na_2SiO_3 activator, to obtain higher mechanical resistance values, a balance between the activator concentration and the curing temperature is necessary. In the durability studies, there was a decrease in mechanical resistance with an increase in exposure cycles to aggressive conditions, and for natural aging, mechanical resistance increased over time.

Author Contributions: Methodology, software, data curation, and formal analysis, I.M.T.B., F.P.d.C. and J.V.F. Conceptualization, supervision, funding acquisition, and project administration, G.d.A.N. and R.R.M. Formal analysis, writing—review and editing, A.M.R. All authors have read and agreed to the published version of the manuscript.

Funding: This research was funded by Coordenação de Aperfeiçoamento de Pessoal de Nível Superior (CAPES), grant number 88887.597478/2021-00; and by Conselho Nacional de Desenvolvimento Científico e Tecnológico (CNPq), grant number 140211/2021-7.

Institutional Review Board Statement: Not applicable.

Informed Consent Statement: Not applicable.

Acknowledgments: The authors would like to thank the support provided by the Laboratório de Tecnologia dos Materiais, where it was possible to carry out the experiments of this work.

Conflicts of Interest: The authors declare no conflict of interest.

References

1. da Silva, A.C.; Méxas, M.P.; Quelhas, O.L.G. Restrictive factors in implementation of clean technologies in red ceramic industries. *J. Clean. Prod.* **2017**, *168*, 441–451. [CrossRef]
2. Azevedo, A.R.G.; França, B.R.; Alexandre, J.; Marvila, M.T.; Zanelato, E.B.; Xavier, G.C. Influence of sintering temperature of a ceramic substrate in mortar adhesion for civil construction. *J. Build. Eng.* **2018**, *19*, 342–348. [CrossRef]
3. Nandi, V.S.; Raupp-Pereira, F.; Montedo, O.R.K.; Oliveira, A.P.N. The use of ceramic sludge and recycled glass to obtain engobes for manufacturing ceramic tiles. *J. Clean. Prod.* **2015**, *86*, 461–470. [CrossRef]
4. Almeida, E.P.; Carreiro, M.E.A.; Rodrigues, A.M.; Ferreira, H.S.; Santana, L.N.L.; Menezes, R.R.; Neves, G.A. A new eco-friendly mass formulation based on industrial mining residues for the manufacture of ceramic tiles. *Ceram. Int.* **2020**, *47*, 11340–11348. [CrossRef]
5. de Figueirêdo, J.M.R.; da Costa, F.P.; Fernandes, J.V.; Rodrigues, A.M.; de Araújo Neves, G.; Menezes, R.R.; de Lima Santana, L.N. Development of Scheelite Tailings-Based Ceramic Formulations with the Potential to Manufacture Porcelain Tiles, Semi-Stoneware and Stoneware. *Materials* **2020**, *13*, 5122. [CrossRef]
6. Fernandes, J.V.; Guedes, D.G.; da Costa, F.P.; Rodrigues, A.M.; Neves, G.d.A.; Menezes, R.R.; Santana, L.N.d.L. Sustainable Ceramic Materials Manufactured from Ceramic Formulations Containing Quartzite and Scheelite Tailings. *Sustainability* **2020**, *12*, 9417. [CrossRef]
7. Siddique, S.; Chaudhary, S.; Shrivastava, S.; Gupta, T. Sustainable utilisation of ceramic waste in concrete: Exposure to adverse conditions. *J. Clean. Prod.* **2018**, *210*, 246–255. [CrossRef]
8. Awoyera, P.O.; Ndambuki, J.M.; Akinmusuru, J.O.; Omole, D.O. Characterization of ceramic waste aggregate concrete. *HBRC J.* **2018**, *14*, 282–287. [CrossRef]
9. Nayana, A.M.; Rakesh, P. Strength and durability study on cement mortar with ceramic waste and micro-silica. *Mater. Today: Proc.* **2018**, *5*, 24780–24791. [CrossRef]
10. Vieira, C.M.F.; Monteiro, S.N. Effect of grog addition on the properties and microstructure of a red ceramic body for brick production. *Constr. Build. Mater.* **2007**, *21*, 1754–1759. [CrossRef]
11. Luhar, I.; Luhar, S.; Abdullah, M.M.A.B.; Nabiałek, M.; Sandu, A.V.; Szmidla, J.; Jurczyńska, A.; Razak, R.A.; Aziz, I.H.A.; Jamil, N.H.; et al. Assessment of the Suitability of Ceramic Waste in Geopolymer Composites: An Appraisal. *Materials* **2021**, *14*, 3279. [CrossRef]
12. Provis, J.; Bernal, S.A. Geopolymers and Related Alkali-Activated Materials. *Annu. Rev. Mater. Res.* **2014**, *44*, 299–327. [CrossRef]
13. Galvão Souza Azevedo, A.; Strecker, K. Kaolin, fly-ash and ceramic waste based alkali-activated materials production by the "one-part" method. *Constr. Build. Mater.* **2020**, *269*, 121306. [CrossRef]
14. Ye, N.; Yang, J.; Liang, S.; Hu, Y.; Hu, J.; Xiao, B.; Huang, Q. Synthesis and strength optimization of one-part geopolymer based on red mud. *Constr. Build. Mater.* **2016**, *111*, 317–325. [CrossRef]
15. Huseien, G.F.; Sam, A.R.M.; Shah, K.W.; Asaad, M.A.; Tahir, M.M.; Mirza, J. Properties of ceramic tile waste based alkali-activated mortars incorporating GBFS and fly ash. *Constr. Build. Mater.* **2019**, *214*, 355–368. [CrossRef]

16. Marvila, M.T.; Alexandre, J.; Azevedo, A.R.G.; Zanelato, E.B.; Xavier, G.C.; Monteiro, S.N. Study on the replacement of the hydrated lime by kaolinitic clay in mortars. *Adv. Appl. Ceram.* **2019**, *118*, 373–380. [CrossRef]
17. Lahoti, M.; Tan, K.H.; Yang, E.-H. A critical review of geopolymer properties for structural fire-resistance applications. *Constr. Build. Mater.* **2019**, *221*, 514–526. [CrossRef]
18. Terrones-Saeta, J.M.; Suárez-Macías, J.; Iglesias-Godino, F.J.; Corpas-Iglesias, F.A. Development of Geopolymers as Substitutes for Traditional Ceramics for Bricks with Chamotte and Biomass Bottom Ash. *Materials* **2021**, *14*, 199. [CrossRef] [PubMed]
19. Hajimohammadi, A.; van Deventer, J.S. Characterisation of One-Part Geopolymer Binders Made from Fly Ash. *Waste Biomass Valori.* **2016**, *8*, 225–233. [CrossRef]
20. Angulo-Ramirez, D.E.; De Gutiérrez, R.M.; Valencia-Saavedra, W.G.; De Medeiros, M.H.F.; Hoppe-Filho, J. Carbonation of hybrid concrete with high blast furnace slag content and its impact on structural steel corrosion. *Mater. Construcción* **2019**, *69*, 182. [CrossRef]
21. Ruiz-Santaquiteria, C.; Fernández-Jimenez, A.; Skibsted, J.; Palomo, A. Clay reactivity: Production of alkali activated cements. *Appl. Clay Sci.* **2013**, *73*, 11–16. [CrossRef]
22. Mehta, A.; Siddique, R. An overview of geopolymers derived from industrial by-products. *Constr. Build. Mater.* **2016**, *127*, 183–198. [CrossRef]
23. Nie, Q.; Hu, W.; Ai, T.; Huang, B.; Shu, X.; He, Q. Strength properties of geopolymers derived from original and desulfurized red mud cured at ambient temperature. *Constr. Build. Mater.* **2016**, *125*, 905–911. [CrossRef]
24. Hajimohammadi, A.; van Deventer, J.S. Solid Reactant-Based Geopolymers from Rice Hull Ash and Sodium Aluminate. *Waste Biomass Valori.* **2016**, *8*, 2131–2140. [CrossRef]
25. Arulrajah, A.; Kua, T.A.; Horpibulsuk, S.; Phetchuay, C.; Suksiripattanapong, C.; Du, Y.J. Strength and microstructure evaluation of recycled glass-fly ash geopolymer as low-carbon masonry units. *Constr. Build. Mater.* **2016**, *114*, 400–406. [CrossRef]
26. Novais, R.M.; Ascensão, G.; Seabra, M.P.; Labrincha, J.A. Waste glass from end-of-life fluorescent lamps as raw material in geopolymers. *Waste Manag.* **2016**, *52*, 245–255. [CrossRef]
27. Mendes, J.P.; Elyseu, F.; Nieves, L.J.J.; Zaccaron, A.; Bernardin, A.M.; Angioletto, E. Synthesis and characterization of geopolymers using clay ceramic waste as source of aluminosilicate. *Sustain. Mater. Technol.* **2021**, *28*, e00264. [CrossRef]
28. Robayo-Salazar, R.A.; Rivera, J.F.; de Gutierrez, R.M. Alkali-activated building materials made with recycled construction and demolition wastes. *Constr. Build. Mater.* **2017**, *149*, 130–138. [CrossRef]
29. Sarkar, M.; Dana, K. Partial replacement of metakaolin with red ceramic waste in geopolymer. *Ceram. Int.* **2020**, *47*, 3473–3483. [CrossRef]
30. Rego, S.R.; Gomes, K.C.; Rosas, M.; Torres, S.M.; de Barros, S. Application of Geopolymeric Adhesives in Ceramic Systems Subjected to Cyclic Temperature Environments. *J. Adhes.* **2013**, *90*, 120–133. [CrossRef]
31. Azevedo, A.R.G.; Vieira, C.M.F.; Ferreira, W.M.; Faria, K.C.P.; Pedroti, L.G.; Mendes, B.C. Potential use of ceramic waste as precursor in the geopolymerization reaction for the production of ceramic roof tiles. *J. Build. Eng.* **2019**, *29*, 101156. [CrossRef]
32. ASTM. *C1437 Standard Test Method for Flow of Hydraulic Cement Mortar*; ASTM International: West Conshohocken, PA, USA, 2020.
33. da Silva, A.L.; Farias, A.F.F.; Pontes, J.R.M.; Rodrigues, A.M.; Costa, A.C.F.d.M. Synthesis of the ZnO-$Ni_{0.5}Zn_{0.5}Fe_2O_4$-Fe_2O_3 magnetic catalyst in pilot-scale by combustion reaction and its application on the biodiesel production process from oil residual. *Arab. J. Chem.* **2020**, *13*, 7665–7679. [CrossRef]
34. Correia, S.L.; Partala, T.; Loch, F.C.; Segadães, A.M. Factorial design used to model the compressive strength of mortars containing recycled rubber. *Compos. Struct.* **2010**, *92*, 2047–2051. [CrossRef]
35. Rovnaník, P. Effect of curing temperature on the development of hard structure of metakaolin-based geopolymer. *Constr. Build. Mater.* **2010**, *24*, 1176–1183. [CrossRef]
36. Medri, V.; Fabbri, S.; Dedecek, J.; Sobalik, Z.; Tvaruzkova, Z.; Vaccari, A. Role of the morphology and the dehydroxylation of metakaolins on geopolymerization. *Appl. Clay Sci.* **2010**, *50*, 538–545. [CrossRef]
37. Al Bakria, A.M.M.; Kamarudin, H.; BinHussain, M.; Nizar, I.K.; Zarina, Y.; Rafiza, A.R. The Effect of Curing Temperature on Physical and Chemical Properties of Geopolymers. *Phys. Procedia* **2011**, *22*, 286–291. [CrossRef]
38. Heah, C.Y.; Kamarudin, H.; Mustafa Al Bakri, A.M.; Binhussain, M.; Luqman, M.; Khairul Nizar, I.; Ruzaidi, C.M.; Liew, Y.M. Effect of Curing Profile on Kaolin-based Geopolymers. *Phys. Procedia* **2011**, *22*, 305–311. [CrossRef]
39. Weng, L.; Sagoe-Crentsil, K. Dissolution processes, hydrolysis and condensation reactions during geopolymer synthesis: Part I—Low Si/Al ratio systems. *J. Mater. Sci.* **2007**, *42*, 2997–3006. [CrossRef]
40. Duxson, P.; Provis, J.L.; Lukey, G.C.; Mallicoat, S.W.; Kriven, W.M.; van Deventer, J.S. Understanding the relationship between geopolymer composition, microstructure and mechanical properties. *Colloids Surf. A Physicochem. Eng. Asp.* **2005**, *269*, 47–58. [CrossRef]
41. Granizo, M.L.; Blanco-Varela, M.T.; Martinez-Ramirez, S. Alkali activation of metakaolins: Parameters affecting mechanical, structural and microstructural properties. *J. Mater. Sci.* **2007**, *42*, 2934–2943. [CrossRef]
42. Criado, M.; Palomo, A.; Fernández-Jimenez, A.M. Alkali activation of fly ashes. Part 1: Effect of curing conditions on the carbonation of the reaction products. *Fuel* **2005**, *84*, 2048–2054. [CrossRef]
43. Singh, J.; Singh, S.P. Geopolymerization of solid waste of non-ferrous metallurgy—A review. *J. Environ. Manag.* **2019**, *251*, 109571. [CrossRef]

44. Narayanan, A.; Shanmugasundaram, P. An Experimental Investigation on Flyash-based Geopolymer Mortar under different curing regime for Thermal Analysis. *Energy Build.* **2017**, *138*, 539–545. [CrossRef]
45. Saludung, A.; Azeyanagi, T.; Ogawa, Y.; Kawai, K. Effect of silica fume on efflorescence formation and alkali leaching of alkali-activated slag. *J. Clean. Prod.* **2021**, *315*, 128210. [CrossRef]
46. Shoaei, P.; Musaeei, H.R.; Mirlohi, F.; Narimani Zamanabadi, S.; Ameri, F.; Bahrami, N. Waste ceramic powder-based geopolymer mortars: Effect of curing temperature and alkaline solution-to-binder ratio. *Constr. Build. Mater.* **2019**, *227*, 116686. [CrossRef]
47. Yousefi Oderji, S.; Chen, B.; Ahmad, M.R.; Shah, S.F.A. Fresh and hardened properties of one-part fly ash-based geopolymer binders cured at room temperature: Effect of slag and alkali activators. *J. Clean. Prod.* **2019**, *225*, 1–10. [CrossRef]
48. Çelikten, S.; Sarıdemir, M.; Özgür Deneme, İ. Mechanical and microstructural properties of alkali-activated slag and slag + fly ash mortars exposed to high temperature. *Constr. Build. Mater.* **2019**, *217*, 50–61. [CrossRef]
49. Nath, P.; Sarker, P. Flexural strength and elastic modulus of ambient-cured blended low-calcium fly ash geopolymer concrete. *Constr. Build. Mater.* **2017**, *130*, 22–31. [CrossRef]
50. Bagheri, A.; Nazari, A.; Sanjayan, J.G. Fibre-reinforced boroaluminosilicate geopolymers: A comparative study. *Ceram. Int.* **2018**, *44*, 16599–16605. [CrossRef]
51. Ganesan, N.; Abraham, R.; Deepa Raj, S. Durability characteristics of steel fibre reinforced geopolymer concrete. *Constr. Build. Mater.* **2015**, *93*, 471–476. [CrossRef]

sustainability

MDPI

Article

Adsorption Behavior of Acid-Treated Brazilian Palygorskite for Cationic and Anionic Dyes Removal from the Water

Vanderlane C. Silva [1], Maria Eduarda B. Araújo [2], Alisson M. Rodrigues [1,2,*], Juliana M. Cartaxo [2], Romualdo R. Menezes [1,2] and Gelmires A. Neves [1,2]

1 Graduate Program in Materials Science and Engineering (PPG-CEMat), Federal University of Campina Grande, Campina Grande, PB 58429-900, Brazil; vanderlanecavalcanti@outlook.com (V.C.S.); romualdo.menezes@ufcg.edu.br (R.R.M.); gelmires.neves@ufcg.edu.br (G.A.N.)

2 Academic Unit of Materials Engineering, Science and Technology Center, Federal University of Campina Grande, Campina Grande, PB 58429-900, Brazil; mariaeduardaba@hotmail.com (M.E.B.A.); julianamelo25@gmail.com (J.M.C.)

* Correspondence: alisson.mendes@professor.ufcg.edu.br

Citation: Silva, V.C.; Araújo, M.E.B.; Rodrigues, A.M.; Cartaxo, J.M.; Menezes, R.R.; Neves, G.A. Adsorption Behavior of Acid-Treated Brazilian Palygorskite for Cationic and Anionic Dyes Removal from the Water. *Sustainability* **2021**, *13*, 3954. https://doi.org/10.3390/su13073954

Academic Editor: P. Senthil Kumar

Received: 11 February 2021
Accepted: 25 March 2021
Published: 2 April 2021

Abstract: The effect of acid treatment on the adsorptive capacity of a Brazilian palygorskite to remove the crystal violet (CV) and congo red (CR) dyes was investigated. The raw palygorskite was acid-treated by different HCl solutions (2, 4, and 6 mol/L). The modifications on the palygorskite structure were investigated using X-ray diffraction, X-ray fluorescence, Fourier-transform infrared spectroscopy, N_2 adsorption/desorption, and thermogravimetric and differential thermal analysis. The efficiency of CV and CR adsorption was investigated, and the effect of the initial concentration, contact time, pH, and adsorbent amount was analyzed. The results revealed that CV adsorption in the acid-treated palygorskite was higher than that of the raw material. A Langmuir isotherm model was observed for the adsorption behavior of CV, while a Freundlich isotherm model was verified for the CR adsorption. A pseudo-second-order model was observed for the adsorption kinetics of both dyes. The higher CV adsorption capacity was observed at basic pH, higher than 97%, and the higher CR removal was observed at acidic pH, higher than 50%. The adsorption parameters of enthalpy (ΔH), entropy (ΔS), and Gibbs energy (ΔG) were evaluated. The adsorption process of the CV and CR dyes on the raw and acid-treated Brazilian palygorskite was predominantly endothermic and occurred spontaneously. The studied raw palygorskite has a mild-adsorption capacity to remove anionic dyes, while acid-treated samples effectively remove cationic dyes.

Keywords: palygorskite; adsorption; anionic dye; cationic dye; water treatment

1. Introduction

The increase in industrial production and the world population has also increased water contamination from industrial and domestic sources [1]. Effluents derived from the textile, food, plastics, and cosmetics industries are rich in dyes [2,3]. Dyes have a complex structure, are stable to light and heat, can be highly toxic, potentially carcinogenic, and mutagenic [4], and rapidly accumulate in living cells and damage human health [5,6].

Crystal violet (CV) is a cationic dye widely used in the textile, plastics, paint, and leather industries [7,8]. The CV dye is not biodegradable; it is toxic, genotoxic, and carcinogenic [5,8,9]. The presence of the crystal violet dye in water can promote the growth of tumors and reproductive abnormalities in fishes [10]. CV dye exposure can cause eye irritation and painful sensitization to light, and in extreme cases, respiratory diseases, kidney failure, chemical cystitis, permanent blindness, and cancer [7,11–13].

Congo red (CR) is an anionic dye often used in the textile, cellulose, rubber, and cosmetics industries [14,15]. In water, besides harming the photosynthesis of plants, the CR dye decomposes into carcinogenic and mutagenic substances [16,17]. In humans, the CR dye can cause skin and eye irritation, nausea, vomiting, and a high level of contamination-induced respiratory diseases and cancer [18,19].

Currently, the removal of dyes from water has become a huge challenge, and efficient and low-cost techniques have been sought with intensity in recent years [9]. Among the developments, nanofiltration [20], photocatalytic oxidation [21], coagulation/flocculation [22], electrochemistry [23], membrane filtration [24] and adsorption [25–27] stand out.

Adsorption presents great effectiveness, ease of operation, and low cost. Adsorbent material can be reused, applied to various pollutants, and be used in water treatment in a broad range of dye concentrations [28–30]. However, some adsorbent materials can present a high cost, especially notable in emerging countries, such as activated carbon [31,32]. This motivated the search for alternative low-cost and efficient adsorbents for the treatment of waters. Clays are often used as dye adsorbents due to their relatively low cost, abundance, availability, non-toxicity, and high adsorptive properties [32–36]. However, one of these clays, the palygorskite, has been studied less than others, despite its high potential for dye adsorption. In recent years, palygorskite has been investigated as an adsorbent for removing various contaminants because of its low cost and high surface area [32,36–40].

An Al-rich palygorskite with an R^{2+}/R^{3+} ratio ($Mg^{2+}/(Al^{3+} + Fe^{3+})$) of $\approx$0.26 from Oman [41] was studied in relation to the removal of cationic/basic dyes, namely, methylene blue (MB) and crystal violet (CV), from aqueous solutions, and the adsorptive capacities of 50.8 mg/g and 57.8 mg/g for MB and CV, respectively, were obtained. Three natural Mg-rich and Al-rich (Mg-poor) palygorskites from China [38] were studied to absorb cationic MB and the adsorptive capacities of 158.03 mg/g, 98.34 mg/g, and 77.92 mg/g were obtained. The palygorskite with the lower R^{2+}/R^{3+} (0.5) ratio and a dioctahedral character had the higher adsorption capacity [38]. However, studies [42] on Algerian palygorskite, with an R^{2+}/R^{3+} ratio of $\approx$0.95 and a dioctahedral-trioctahedral character, obtained an adsorptive capacity of just 15 mg/g for MB. These results indicate that the dioctahedral character, Al-rich palygorskite seems to favor the adsorption of cationic dyes.

An investigation that used a palygorskite from the Unites States with an R^{2+}/R^{3+} of $\approx$0.95 for the removal of reactive red 120 (anionic dye) [43] observed a low adsorptive capacity, not exceeding 0.5 mg/g. A Brazilian palygorskite [44], with an R^{2+}/R^{3+} of 0.75 and a dioctahedral-trioctahedral character, was used to remove remazol yellow (anionic dye) and obtained an adsorptive capacity of 20.08 mg/g. These studies indicate that palygorskites with high R^{2+}/R^{3+} present a low adsorption capacity for anionic dyes.

Yang et al. [33] investigated the adsorption capacity of the anionic and cationic dyes on the commercial Chinese palygorskite, which some metal ions (Ca^{2+}, Mg^{2+}, Na^{+}, and K^{+}) were removed with the aid of soxhlet extraction. The metal ions removed implied more adsorption active sites for cationic dyes; thus, there was an improvement to the adsorption capacity for these dyes, such as the crystal violet and cationic light yellow dyes. On the other hand, before removing the metal ions, the commercial Chinese palygorskite showed a relatively high adsorptive capacity for anionic methyl orange dye. In this case, the metal ions act as bridges for efficient adsorption for these dyes. Other factors, such as the molecular structure of the dyes, were also identified as efficient in the adsorption. Another Chinese palygorskite was acid-treated with HCl [45] to be used in the adsorption of MB, and a high adsorption capacity (216.5 mg/g) was achieved. These studies indicate that acid treatment improves adsorption capacity for cationic and anionic dyes in palygorskites with a dioctahedral-trioctahedral character.

These studies indicate a high potential for water treatment with palygorskites. It was observed that the adsorptive potential of Chinese palygorskites has been well discussed in the literature. In contrast, few studies have approached the adsorptive potential of the Brazilian palygorskite for dye removal [44,46]. Furthermore, little has been explored about how acid treatment affects the Brazilian palygorskite structure [47]. Thus, this work aimed to analyze the effect of acid treatment on Brazilian palygorskite structure and to evaluate its potential in treating water contaminated with dyes.

2. Materials and Methods

2.1. Raw Materials

Palygorskite clay investigated in this work it is from Guadalupe City, Piauí State, Brazil, and was supplied by União Brasileira de Mineração S.A. (UBM). The crystal violet dye (CV) (Synth, Brazil), congo red dye (CR) (Dinâmica Química, Brazil), hydrochloric acid, and ammonium hydroxide (VETEC, Brazil) were the other raw materials used in this study.

2.2. Acid Treatment

The as-received palygorskite was treated at 60 °C for 24 h, sieved (0.074 mm), and acid-treated with HCl. Then, 50 g of palygorskite was immersed in 500 mL of hydrochloric acid solution in different concentrations (2, 4, and 6 mol/L) under reflux conditions for 2 h at 70 °C. The mixture was filtered under vacuum and washed several times with distilled water until the filtered solution reached a pH = 7. Finally, the material obtained was dried for 12 h at 60 °C. Table 1 shows the nomenclatures used for the raw and activated palygorskite.

Table 1. Nomenclature of the samples with and without acid treatment.

Treatment	Acid Concentration	Sample Nomenclature
None	-	Pal
Acid	HCl–2 mol/L	Pal-2M
Acid	HCl–4 mol/L	Pal-4M
Acid	HCl–6 mol/L	Pal-6M

2.3. Characterizations

X-ray diffraction (Shimadzu, XRD-6000) was performed using CuKα (λ = 1.54 Å), operated at 40 kV and 30 mA, in a 2θ angular range of 5°–50° and 0.02° of step size [46,47]. The chemical analysis was determined using an X-ray fluorescence spectrometry (Shimadzu, EDX-720). Infrared spectra with Fourier-transform infrared spectrometry (FTIR) were recorded in the spectral range from 4000 cm^{-1} to 400 cm^{-1}, with 32 scans and 4 cm^{-1} resolution, using KBr pellets (Bruker, Vertex-70). Thermogravimetric (TG) and differential thermal analysis (DTA) were performed under an air atmosphere, with a heating rate of 10 °C/min (Shimadzu, DTG-60H). A nitrogen adsorption/desorption test (Quantachrome, Autosorb iQ) was performed and the specific surface area and pore diameter were determined using the Brunauer–Emmett–Teller (BET) method [48,49].

2.4. Adsorption Experiments

Adsorption tests were performed to investigate the effect of contact time, initial adsorbate concentration, adsorbent dosage, and pH variation. The adsorbent-solution systems were shaking (150 rpm) at 25 °C for up 360 min. After shaking, the samples were centrifuged at 3600 rpm for 5 min. The residual dye concentrations in the supernatant were measured by absorbance at 582.5 nm and 501 nm for λ_{max} for the CV and CR dyes, respectively. The experiments were accomplished by UV-vis absorption spectrophotometry (UV-1800, Shimadzu). The amount of CV and CR dyes adsorbed on the Pal, Pal-2M, Pal-4M, and Pal-6M samples was calculated using Equation (1) [50], and the removal percentage was calculated using Equation (2) [51]:

$$q_e = [(C_o - C_e)V]/m \quad (1)$$

$$\%adsorbed = [(C_o - C_e)/C_o] \times 100 \quad (2)$$

where q_e (mg/g) is the adsorption capacity, C_o (mg/L) and C_e (mg/L) are the initial and equilibrium concentrations, respectively, V (L) is the volume of the solution, and m (g) is the mass of the palygorskite samples.

To study the effect of the CV and CR dye concentrations, 20 mg of the investigated samples were added in flasks containing 20 mL of the dye solution with different concentrations ranging from 2.5 to 200 mg/L. The pH of 7 was kept constant. The effects of contact time (15–360 min), pH (3, 5, 7, 9, and 11), and temperature (25–55 °C) were investigated by adding 20 mg of the investigated samples (Pal, Pal-2M, Pal-4M, and Pal-6M) into 20 mL of the CV or CR solution with a 50 mg/L concentration. The initial pH of the solutions was adjusted by adding 0.1 M NH_4OH or HCl. The influence of the adsorbent dosage was analyzed by varying the amount of samples between 10 to 40 mg at a concentration of CV and CR of 50 mg/L at pH = 7.

2.5. Isothermal, Kinetic, and Thermodynamic Studies

The Langmuir (Equation (3)) [52] and Freundlich (Equation (4)) [53,54] isothermal models were used in the analysis of adsorption data:

$$C_e/q_e = [1/(q_{max} \cdot K_L)] + (C_e/q_{max}) \quad (3)$$

where q_e (mg/g) is the amount adsorbed in the equilibrium, C_e(mg/L) is the equilibrium concentration, q_{max} (mg/g) represents the maximum adsorption, and K_L (L/mg) is the Langmuir constant.

$$\log q_e = \log K_f + 1/n\ (\log C_e) \quad (4)$$

where C_e and q_e are as defined above, and K_f ((mg/g)(L/mg)$^{1/n}$) and 1/n are the Freundlich constants characteristic of the system. K_f represents an approximate value of the adsorption capacity, and 1/n indicates the intensity of the adsorption process and is related to the heterogeneity of the surface, with values between 0 and 1; the closer the value is to 0, the more heterogeneous the surface.

To examine the adsorption kinetics, the pseudo-first-order (Equation (5)) [55] and pseudo-second-order (Equation (6)) [56] models were used:

$$\ln(q_e - q_t) = \ln q_e - k_1 t \quad (5)$$

where q_e is as defined above, q_t (mg/g) is the amount of adsorbate adsorbed at a given time, k_1 (min^{-1}) is the pseudo-first-order speed constant, and *t* is the contact time in minutes (min).

$$t/q_t = 1/(k_2 q_e^2) + t/q_e \quad (6)$$

where q_e and q_t are the same as defined above, k_2 (g.mg^{-1}·min^{-1}) is the pseudo-second-order speed constant, and *t* is the contact time in minutes (min).

The thermodynamic parameters were determined from the experimental data using the Van't Hoff equations (Equations (7)–(9)) [42,57]:

$$\mathrm{Ln}\ K_d = (\Delta S/R) - (\Delta H/RT) \quad (7)$$

$$\Delta G = -\ RT\mathrm{Ln}\ K_d \quad (8)$$

$$K_d = q_e/C_e \quad (9)$$

where R (8.314 J/mol K) is the universal gas constant, T (K) is the studied temperature, K_d is the distribution coefficient for adsorption, and ΔH, ΔS, and ΔG are the variations of enthalpy, entropy, and Gibbs energy, respectively.

2.6. Regeneration of Adsorbents

The adsorbents loaded with CV and CR were desorbed using 40 mL of 1 M CH_3COOH solution and stirred in a shaker at 120 rpm for 8 h. Then, the adsorbents were washed with distilled water and placed in the oven at 60 ° C for 12 h. After drying, they were reused for 4 cycles.

3. Results and Discussion

3.1. Characterization of Raw and Acid-Treated Palygorskite

X-ray diffraction (XRD) patterns of the raw and acid-treated (2, 4, and 6 mol/L) Brazilian palygorskite are illustrated in Figure 1. Palygorskite (ICCD 21-0958) and quartz (ICCD 46-1045) phases were identified in the raw and acid-activated clays, as observed in other studies [58–60]. Câmara et al. and Suarez et al. [39,58] reported that the palygorskite can be classified into three types: Type I has similar Al and Mg contents and negligible isomorphic substitutions ($d_{200} < 6.35$ Å); Type II is rich in Al contents and has a dioctahedral character, and this is the most commonly found type ($6.35 < d_{200} < 6.46$ Å); and Type III is rich in Mg contents and has a trioctahedral character ($d_{200} > 6.46$ Å). The Brazilian palygorskite investigated in this work has a d_{200} equal to 6.39 Å; therefore, it belongs to Type II.

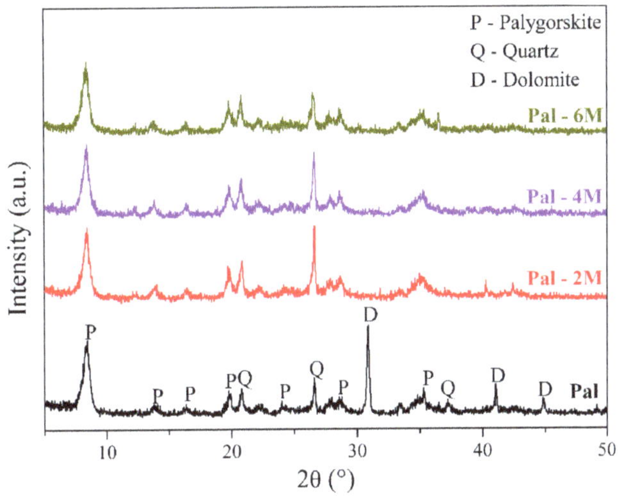

Figure 1. X-ray diffraction (XRD) pattern of raw palygorskite and acid-treated palygorskite.

The XRD patterns indicated that the structure of the Brazilian palygorskite was not significantly altered by the acid treatment. Quartz reflections remained stable due to their insolubility in acidic solution [61–63]. Studies depicted the formation of amorphous SiO_2 because of the acid treatment [61,63,64]; however, no amorphous band [65,66] was identified in the range of 2θ between 20 and 35° in the acid-treated Brazilian palygorskite. The absence of an amorphous band can be attributed to the short acid-treatment time, which was inefficient in destroying the tetrahedral crystalline sheet of palygorskite. On the other hand, the reflections corresponding to dolomite (ICCD 36-0426) were not identified in the XRD standards of the acid-treated samples, indicating the dissolution of this carbonate with the acid treatment observed in other studies [64].

The chemical analysis (Table 2) showed that the palygorskite samples investigated were mostly composed of SiO_2, MgO, and Al_2O_3, which indicates the presence of hydrated magnesium and aluminum silicate on this palygorskite mineral [61]. The raw palygorskite had an R^{2+}/R^{3+} ratio $\approx$1, but the acid-treated samples had an R^{2+}/R^{3+} ratio $\approx$0.6. The R^{2+}/R^{3+} ratio $\approx$1 is related to the high amount of MgO in the sample due to dolomite presence. After acid-treatment, dolomite was removed; thus, the more confident R^{2+}/R^{3+} ratio $\approx$ 0.6 was obtained. This 0.6 ratio indicates that the studied sample had an intermediate dioctahedral and trioctahedral character, such as the other Brazilian palygorskite already studied [44]. Palygorskite with such characteristics [44] presented a low adsorption capacity for anionic dye.

Table 2. Chemical composition (wt%) of raw and acid-treated palygorskite.

Sample	SiO_2	MgO	Al_2O_3	CaO	Fe_2O_3	K_2O	Other Oxides
Pal	52.8	13.9	13.5	11.9	5.3	0.9	1.7
Pal-2M	69.3	8.3	13.5	0	5.2	0.9	2.8
Pal-4M	71.9	7.6	13.4	0	4.5	0.8	1.8
Pal-6M	73.4	7.1	13.4	0	4.0	0.8	1.2

The acid-treated samples (Pal-2M, Pal-4M, and Pal-6M) presented a progressive decrease in the MgO, Al_2O_3, and Fe_2O_3 contents (octahedral cations) with the rise in HCl concentration. The decrease in the MgO, Al_2O_3, and Fe_2O_3 contents indicates that there was leaching of Mg^{2+}, Al^{3+}, and Fe^{3+} ions in the treated samples because of the dissolution of dolomite (contain MgO) [67] and the removal of ions of the octahedral sheet.

The extent of the leaching of Mg^{2+} and Fe^{3+} ions was higher than that of Al^{3+} ions. According to the literature [32,62,68], the Mg^{2+} and Fe^{3+} cations are preferably located on the edges of the octahedral sheets. Therefore, the Mg^{2+} and Fe^{3+} cations are the first ones attacked by the H^+ protons, which results in the most significant leaching [32,62,68]. The higher leaching of octahedral sheet cations resulted in a more exposed silica sheet after acid-treatment [63,64], which can influence the adsorption behavior of the final material.

After acid-treatment, the CaO content was reduced to zero, since the dolomite was dissolved in an acidic medium [67], corroborating the XRD results. The decrease in K_2O content is due to the exchangeable cations being replaced by H^+ ions during the acid treatment [69,70].

The FTIR spectra data measured from the raw and acid-treatment Brazilian palygorskite are shown in Figure 2. From the raw palygorskite, the observed 3612 cm^{-1} band can be attributed to the Al-OH-Al stretching vibration and the 3578 cm^{-1} band was associated with the Al-Fe^{3+}-OH or Al-Mg-OH stretching vibrations [64,71–73]. The 3541, 3374, 3268, and 1656 cm^{-1} bands were also detected and related to the zeolitic and coordinate waters present in the palygorskite structure [42,74]. The calcite was related to bands that appeared in 1439 and 729 cm^{-1} [75,76], and it became evident that there were carbonate impurities in the studied sample. The bands between 1190 and 975 cm^{-1} corresponded to Si-O bond length. The 1190 and 642 cm^{-1} bands were considered palygorskite fingerprints and referred to the asymmetric and symmetrical elongation of the Si-O-Si connections [77,78]. The 909 cm^{-1} band can be attributed to the Al-OH-Al deformation and was considered a consequence of the palygorskite's dioctahedral character [38,76]. The 877 cm^{-1} band was related to the flexion vibration mode of the Al-Fe-OH bond [38]. The 580 cm^{-1} band corresponded to the Si-O deformation vibration [77,79].

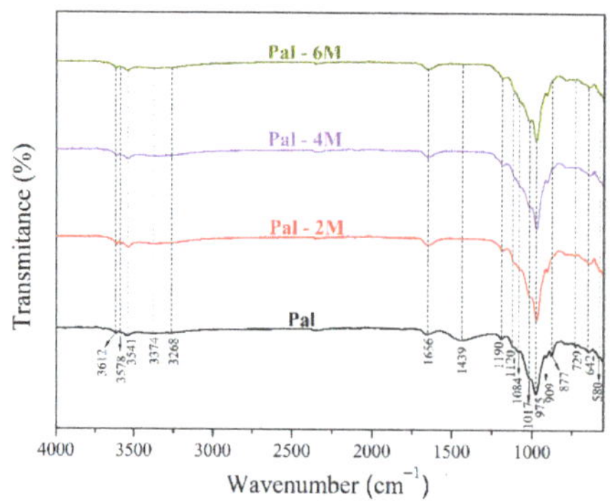

Figure 2. Fourier-transform infrared spectrometry (FTIR) spectra of the raw and acid-treated palygorskite.

In Figure 2, it was observed that after acid-treatment, the bands referring to calcite (1439 and 729 cm^{-1}) and the 877 cm^{-1} band disappeared, while the intensities of the 1190 and 975 cm^{-1} bands increased. This was related to dolomite dissolution and Fe leaching from the octahedral sheet. Octahedral cations are generally more susceptible to leaching than the tetrahedral Si [80,81], which favors the intensity of Si-related bands.

Some mass-loss events were identified on TG analysis (Figure 3). The first event between 23–125 °C was related to the evaporation of water that was physically adsorbed [82], and the values measured were equal to 6.5%, 9.4%, 8.4%, and 9.5% for the Pal, Pal-2M, Pal-4M, and Pal-6M samples, respectively. The second event (125–230 °C) corresponded to the coordinated and zeolitic water molecules located in the palygorskite's channels [83]. The percentages of mass loss measured for the second event were 2.3%, 3.4%, 2.5%, and 2.4% for the Pal, Pal-2M, Pal-4M, and Pal-6M samples, respectively. The third event occurred between 230–530 °C and presented a mass loss equal to 4.9%, 5.3%, 5.4%, and 5.1% to the Pal, Pal-2M, Pal-4M, and Pal-6M samples, respectively. These mass losses can be attributed to the dehydroxylation process and condensation of silanol and aluminol groups [39,84]. Only the Pal sample showed a mass loss (12.6%) between 530–720 °C. This thermal event is related to dolomite decomposition [85]. The dolomite composition agreed with the aforementioned chemical analysis, which showed that the acid-treatment released CaO contents from the Brazilian palygorskite. The DTA curves presented endothermic peaks in the same temperature range of the losses of mass observed in the TG curves. The DTA peaks were attributed to the evaporation of physically adsorbed water molecules [86], the release of zeolitic water molecules [87], the removal of coordinated water, the condensation of surface groups [88], and the dolomite decomposition [89].

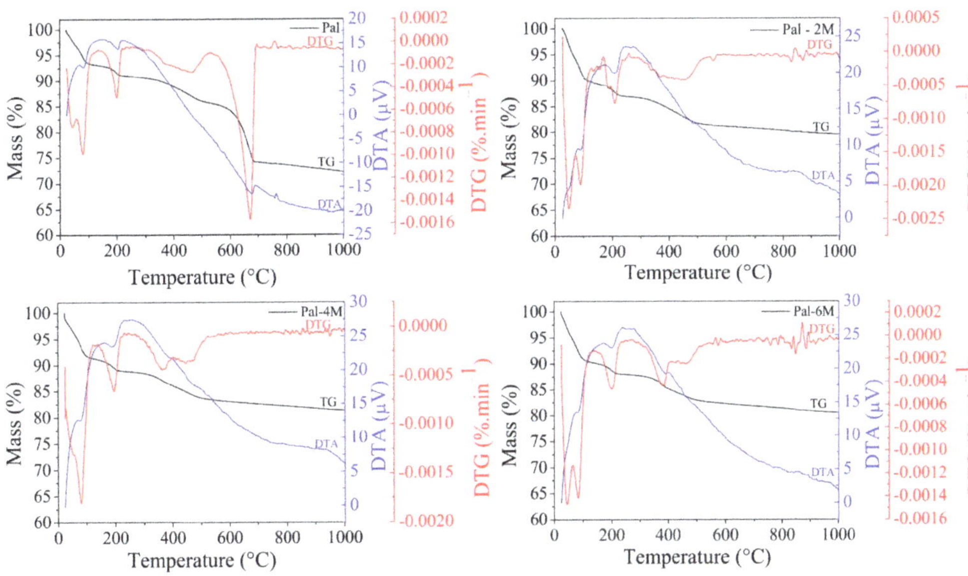

Figure 3. Thermogravimetric (TG) and differential thermal analysis (DTA) curves of the raw and acid-treated palygorskite.

The acid-treated samples presented endothermic peaks similar to those observed in the raw palygorskite, except the peak related to the loss of mass of the dolomite decomposition, which was not observed in the acid-treated samples because of the removal of this impurity. The 4 M and 6 M acid-treated samples also presented a different loss of mass behavior between 350 °C and 500 °C, with two peaks in TG instead of a broad band of mass loss, as was observed in the raw and the 2 M acid-treated sample. This can be related to alterations in the dehydroxylation kinetic process.

The specific surface area (S_{BET}) and average pore diameter (D_p) of the raw and acid-treated Brazilian palygorskite are listed in Table 3. After acid-treatment, the increase in S_{BET} can be explained because of the removal of carbonate impurities, leaching of cations from the octahedral sheet, and disaggregation of particles [80,90]. However, when the acid concentration increased from 4 M to 6 M, a small S_{BET} reduction was observed. This behavior can be related to a more pronounced collapse of the palygorskite channels [32,64]. Also, a decrease in D_p values was observed for the Pal-4M and Pal-6M samples, possibly because, in high HCl concentrations, there is a partial collapse of the channels and a decrease in pore size. However, Gonzalez et al. [91] reported developing a new pore system with a smaller pore size in acid-treated palygorskite. This indicates that with the rise in acid concentration, there is an increase in surface area due to a leaching of octahedral sheets and the exposition of new reactive sites. At high acid concentration, the channels begin to collapse, decreasing the average pore size.

Table 3. Specific surface area (S_{BET}) and average pore diameter (D_p) of the raw and acid-treated palygorskite.

Sample	Specific Surface Area (m^2/g)	Average Pore Diameter (nm)
Pal	80.4	14.3
Pal-2M	141.2	15.2
Pal-4M	182.2	13.0
Pal-6M	176.9	13.3

3.2. Optimization of Adsorption Conditions

3.2.1. Effect of Dye Concentrations

The effect of the initial concentration of crystal violet (C_i^{CV}) and congo red (C_i^{CR}) dyes on the adsorption capacity of raw and acid-treated Brazilian palygorskite is shown in Figure 4a,b. Different CV and CR concentrations (2.5 to 200 mg/L) were tested for a contact time equal to 360 min, pH 7, and clay samples with a weight of 20 mg. It was observed that the adsorbed amount (q_e) raised significantly as C_i^{CV} and C_i^{CR} increased. This behavior is explained in terms of the greater probability of collision between the dye molecules and the palygorskite surface. Besides, the increases in C_i^{CV} and C_i^{CR} contribute to increasing the driving force at the solid–liquid interface that exceeds the mass transfer resistance, implying an increase in the adsorption capacity [28].

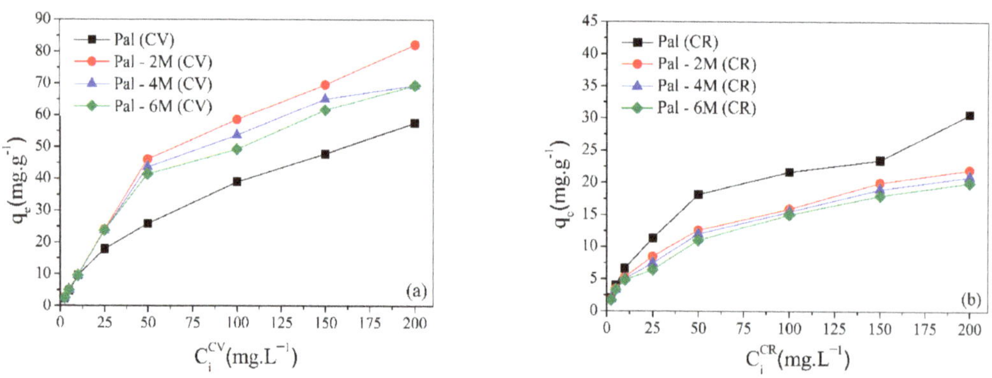

Figure 4. Effect of initial concentration on the adsorption capacity of raw and acid-treated palygorskite for (**a**) CV and (**b**) CR.

On the raw Brazilian palygorskite, the adsorption capacity values of the CV dye were higher than those measured for the CR dye. Such behavior can be explained in terms of the electrostatic attraction, since that raw palygorskite contains negative charges in its surface [30,33]. The maximum CV adsorption capacity was: Pal-2M (82.1 mg/g) > Pal-4M (69.4 mg/g) > Pal-6M (69.3 mg/g) > Pal (57.5 mg/g). Compared to the raw palygorskite, the amount adsorbed increased by 43%, 21%, and 20.5% for the Pal-2M, Pal-4M, and Pal-6M samples, respectively. This indicates that the adsorption performance of palygorskite for CV was improved with the acid-treatment. The amount adsorbed by raw clay was similar to the amount adsorbed by clay with $R^{2+}/R^{3+} \approx 0.26$ [41], with a dioctahedral character. This indicates that clays with dioctahedral and dioctahedral-trioctahedral characters (as in the case of this work) have a similar CV adsorption capacity.

The maximum CR adsorption capacity was: Pal (30.6 mg/g) > Pal-2M (22 mg/g) > Pal-4M (20.9 mg/g) ≈ Pal-6M (20 mg/g). The lower adsorption capacity of the palygorskite to the CR was attributed to the electrostatic repulsion between the negatively charged surface and the negative nature of the CR dye [92]. Furthermore, Yang et al. [33] pointed out that the leaching of Mg^{2+} and Ca^{2+} ions decreases the zeta potential of palygorskite clay and reduces the anionic dyes' absorptive capacity. This behavior can be explained in terms of the exposition of silica negatively charged sites, which improved the adsorption of the cationic CV dye, but reduced the CR adsorption capacity after acid treatment.

3.2.2. Effect of Contact Time

Figure 5a,b shows the amount of CV and CR adsorbed on the raw and acid-treated samples as a function of contact time. CV and CR solutions with initial concentrations of 50 mg/L, for a constant 20 mg clay sample mass and pH 7 were used to study the adsorption kinetics. Initially, the CV adsorption was fast for all samples, achieving about 40% of adsorption in the first 30 min. This is due to the numerous active sites available on the surface of the adsorbent in the early stages [45] and the strong electrostatic interaction between the negatively charged palygorskite surface and the cationic CV [41]. When equilibrium was reached in 240 min, the adsorption capacity and CV removal values were 26.02 mg/g and 52%, respectively. The Pal-2M, Pal-4M, and Pal-6M samples removed approximately 92% (45.95 mg/g), 84% (42.1 mg/g), and 80% (40.38 mg/g) of the CV, respectively. For the CR dye, the adsorption kinetics was slower, and in 30 min, less than 18% of the dye was removed. The equilibrium was achieved in 240 min with the removal of 36.34% (18.17 mg/g), 25.12% (12.56 mg/g), 23.98% (11.99 mg/g), and 22% (11 mg/g) of the CR dye for the Pal, Pal-2M, Pal-4M, and Pal-6M samples, respectively.

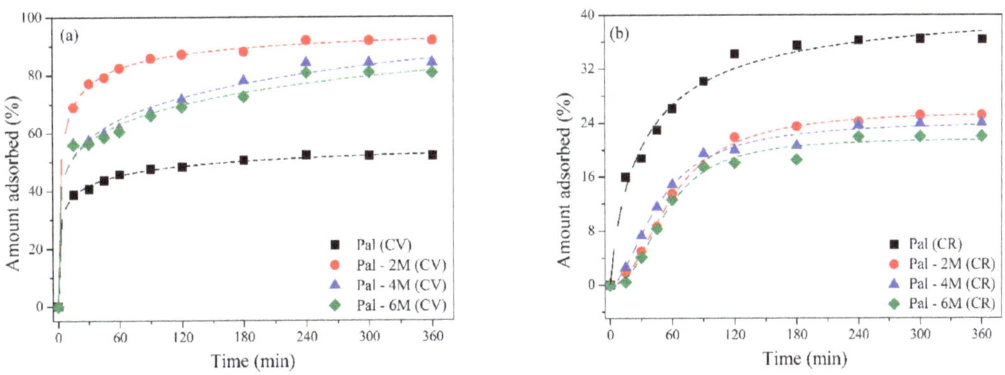

Figure 5. Effect of contact time on (**a**) crystal violet (CV) adsorption and (**b**) congo red (CR) adsorption on palygorskite before and after acid treatment.

3.2.3. Effect of pH Variation

Figure 6a,b shows the effects of pH on the adsorption of CV and CR in the palygorskite before and after acid treatment. The experiments were performed using 50 mg/L of the initial concentration of CV and CR solutions with the pH values from 3 to 11 for a contact time of 360 min. The amount of adsorbed CV and CR depends on the initial pH of the solution. As the pH increased from 3 to 11, the adsorbed amount of CV rose, while CR decreased. In acidic media, the H^+ ions present in the solution protonate the palygorskite surface and the functional groups (–Si-OH) on its surface to –Si-OH^{2+}. The competition between H^+ ions and the CV cationic dye for the adsorption sites, and the repulsive force between the positively charged adsorbent surface and CV molecules, are considered the factors responsible for the lower adsorption rate of CV at a low pH [93].

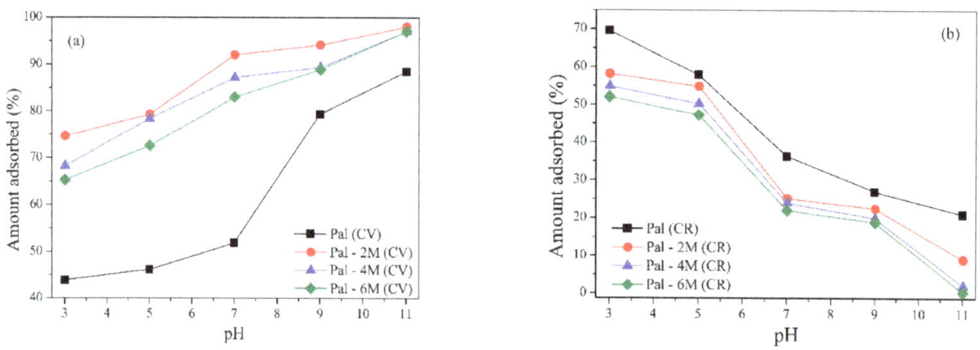

Figure 6. Effect of pH on (**a**) CV adsorption and (**b**) CR adsorption in palygorskite before and after acid treatment.

On the other hand, negatively charged CR molecules interacted electrostatically with the positively charged surface of the adsorbent at low pH, presenting a higher rate of adsorption [94]. In basic pH, OH^- ions promote the deprotonation of the surface and generate –Si-O^- groups, which favor the adsorption of the positively charged CV [95]. In addition to competing with OH^- ions in basic pH, CR molecules are also repelled by the negative surface of the adsorbent, which decreases the adsorption of this anionic dye [96]. At pH 11, the acid-treated samples showed a removal higher than 97% for CV. At pH 3, the raw and acid-treated samples provided a removal higher than 50% for CR.

3.2.4. Effect of Adsorbent Dosage

Different amounts of adsorbents (10 to 40 mg) were used for the adsorption of 50 mg/L of CV and CR. The experiments were performed at pH 7 for 360 min (see Figure 7a,b). When a small amount of palygorskite was used, the adsorbed amount increased considerably when the adsorbent mass increased from 10 to 40 mg. This can be attributed to the increase in the adsorption sites available to remove CV and CR [40,97]. With 40 mg of adsorbent, the CV removal rate reached 82.5%, 97.6%, 96.9%, and 96.6% for the Pal, Pal-2M, Pal-4M, and Pal-6M samples, respectively. For the CR, the removal rate was 61.1%, 44.1%, 30.1%, and 29.2% for the Pal, Pal-2M, Pal-4M, and Pal-6M samples, respectively.

In Figure 7a, more precisely for the Pal sample, an increase of 30% was observed on the amount adsorbed for the adsorbent dosages between 20 and 30 mg. Such behavior occurs because the carbonate presence probably causes the formation of the agglomerates on the adsorbent. These agglomerates decrease the adsorption efficiency, because a greater amount of interactions between the adsorbate and the absorbent are required for the adsorption to be effective.

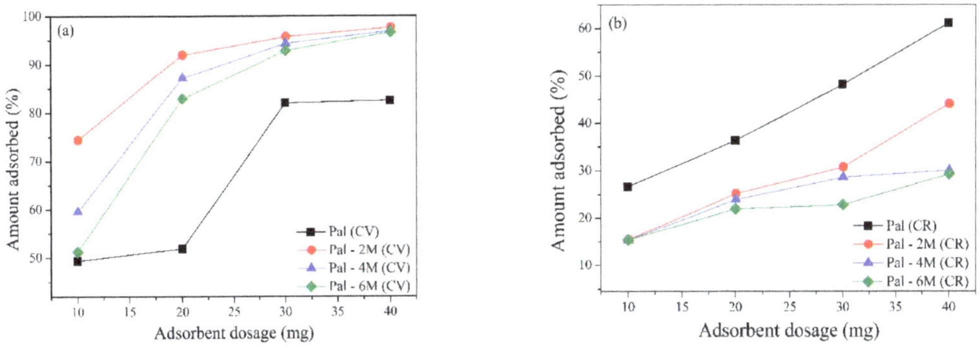

Figure 7. Effect of adsorbent dosage on (**a**) CV adsorption and (**b**) CR adsorption in palygorskite before and after acid treatment.

3.2.5. Effect of Temperature

Figure 8 shows the temperature effect on the CV and CR removal at a temperature range between 25–55 °C. For this, CV and CR solutions with initial concentrations of 50 mg/L and a 20 mg clay sample mass were used under pH = 7. The adsorbed amounts of CV and CR increased slightly with temperature. Such behavior occurs because, at higher temperatures, the kinetic energy of the dye molecules increases, which in turn increases the collision rate of the dye and the clay [98]. Furthermore, the results imply that the CV and CR adsorptions in the Pal, Pal-2M, Pal-4M, and Pal-6M samples are a predominantly endothermic process, which means that the amount of dye adsorbed increases with higher temperatures [99].

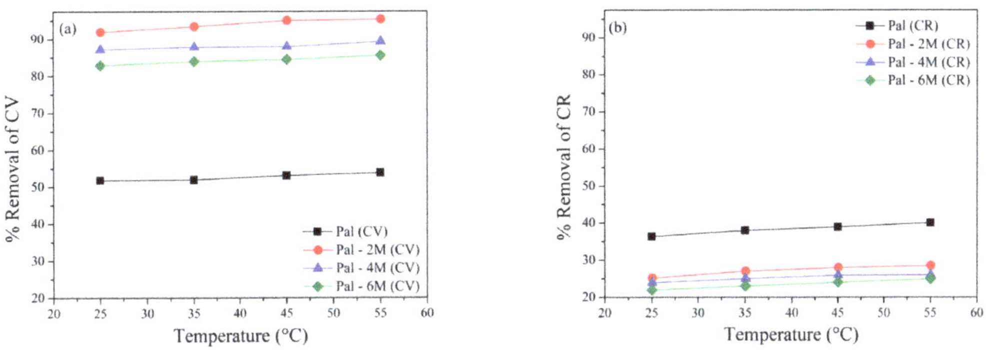

Figure 8. Effect of temperature on (**a**) CV adsorption and (**b**) CR adsorption in palygorskite before and after acid treatment.

3.3. Adsorption Isotherms

The isothermal adsorption data were adjusted to the Langmuir and Freundlich models (Figure 9a,d and Table 4). Through the values of the correlation coefficients (R^2) and the errors, it was observed that the Langmuir model better described the CV adsorption process. In contrast, the CR adsorption was better adjusted to the Freundlich model. The best fit of the experimental data of CV to the Langmuir model suggests that their adsorption occurs mainly via the chemisorption process, with monolayer adsorption on the palygorskite surface [39,52,100]. This was not the case for CR adsorption, where Freundlich was a much better fit. This indicates that CR adsorption on the raw and acid-treated palygorskite occurs on a heterogeneous surface with the formation of multiple layers [101,102]. These data agree with a negative palygorskite surface that favors the chemisorption of cationic dyes, such as CV, and the interaction in specific positively charged sites of anionic dyes, such as CR, results in heterogeneous adsorption.

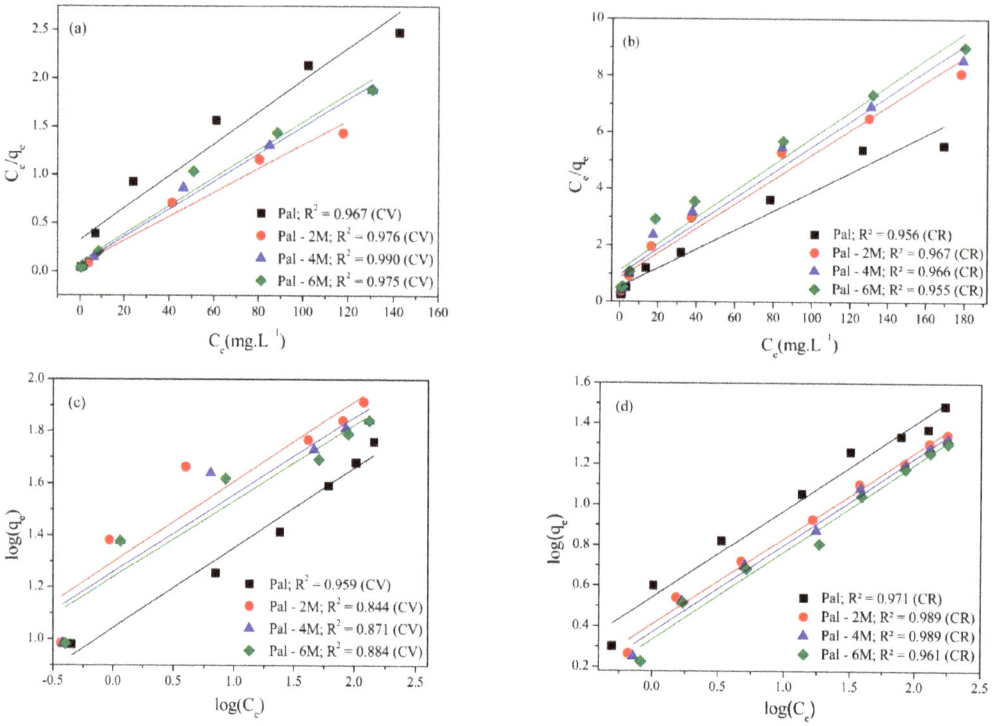

Figure 9. Linear adjustment for CV adsorption and CR adsorption data according to the (**a**,**b**) Langmuir and (**c**,**d**) Freundlich models.

Table 4. Parameters of Langmuir and Freundlich isothermal models for CV adsorption and CR adsorption.

	CV				CR			
Models	**Pal**	**Pal-2M**	**Pal-4M**	**Pal-6M**	**Pal**	**Pal-2M**	**Pal-4M**	**Pal-6M**
Langmuir								
q_{max}(mg·g^{-1})	60.2	83.3	69.9	69.0	30.3	25.3	22.2	21.3
K_L (L·mg^{-1})	0.05	0.20	0.17	0.15	0.06	0.04	0.03	0.03
R_L	0.09	0.03	0.02	0.03	0.08	0.11	0.14	0.14
R^2	0.967	0.976	0.990	0.975	0.956	0.967	0.966	0.955
Error	0.07	0.04	0.02	0.05	1.31	1.75	1.99	2.83
Freundlich								
1/n	031	0.31	0.30	0.30	0.43	0.44	0.43	0.43
K_f (mg$^{1-1/n}$.Kg^{-1}.L$^{1/n}$)	11.0	19.8	18.1	17.3	3.48	2.75	2.70	2.69
R^2	0.959	0.844	0.871	0.884	0.971	0.989	0.989	0.961
Error	0.09	0.08	0.06	0.06	0.03	0.01	0.01	0.02

3.4. Adsorption Kinetics

Figure 10a,d and Table 5 show the results obtained from adjusting the kinetic data of CV and CR adsorption to the pseudo-first-order and pseudo-second-order models. The experimental data were better adjusted to the pseudo-second-order model because the R^2 values were the closest to 1 ($R^2 > 0.990$). The calculated q_e agreed with the experimental data

(q_{exp}). These results indicate that adsorption in palygorskite samples can be determined mainly by chemisorption, involving the sharing or transfer of electrons [103,104].

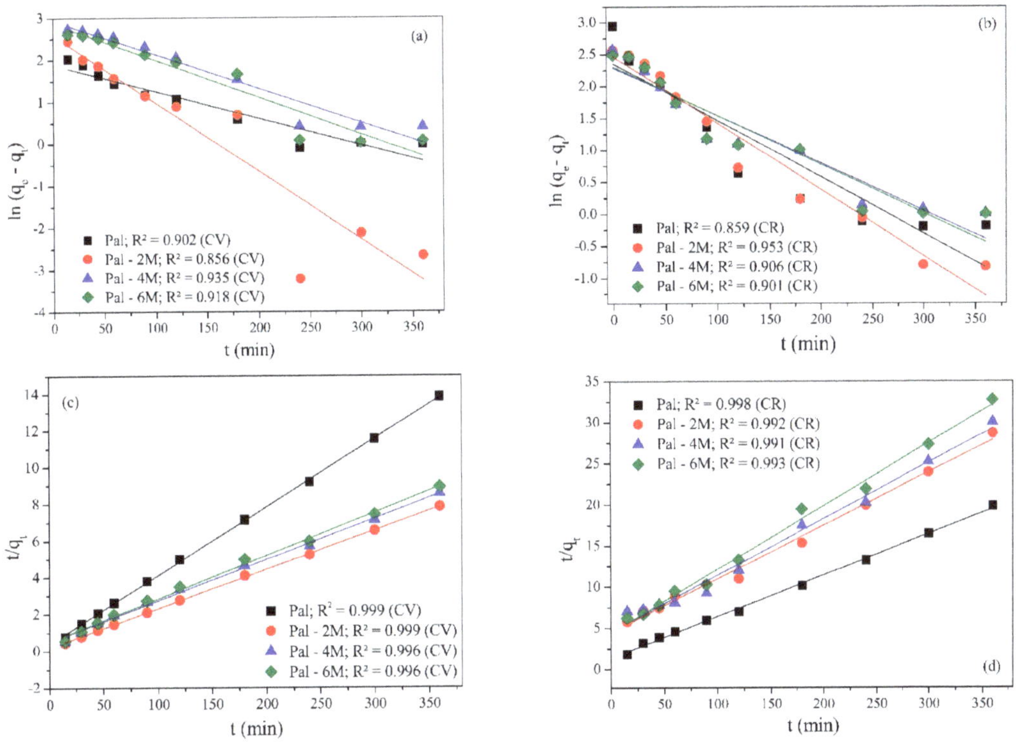

Figure 10. Linear adjustment of (**a**,**b**) pseudo-first-order and (**c**,**d**) pseudo-second-order kinetic models for CV adsorption and CR adsorption.

Table 5. Kinetic parameters of pseudo-first-order and pseudo-second-order models for CV adsorption and CR adsorption.

	CV				CR			
Models	**Pal**	**Pal-2M**	**Pal-4M**	**Pal-6M**	**Pal**	**Pal-2M**	**Pal-4M**	**Pal-6M**
Pseudo-first-order								
$q_{e,exp}$ (mg.g^{-1})	25.97	45.98	42.08	40.36	18.17	12.56	11.99	11.01
$q_{e,cal}$ (mg.g^{-1})	77.31	194.17	126.30	112.06	126.46	208.40	190.55	199.52
k_1 (min^{-1})	0.02	0.04	0.02	0.02	0.02	0.02	0.01	0.02
R^2	0.902	0.856	0.935	0.918	0.859	0.953	0.906	0.901
Error	0.51	5.17	0.53	0.81	2.67	3.96	7.26	4.66
Pseudo-second-order								
$q_{e,cal}$ (mg.g^{-1})	26.76	46.95	44.62	42.52	20.02	15.46	14.71	13.01
k_2 (min^{-1})	3.87	2.74	1.05	1.17	0.04	0.02	0.02	0.02
R^2	0.999	0.999	0.996	0.996	0.998	0.992	0.991	0.993
Error	0.07	0.02	0.21	0.27	1.70	0.72	0.77	0.86

3.5. Adsorption Thermodynamics

The ΔH and ΔS values were calculated from the slope and intercept of the ln K_d versus 1/T plot (Figure 11a,b). The ΔH, ΔS, and ΔG (Equation (8)) values are listed in Tables 6 and 7. Adsorption processes with positive ΔH values are predominantly endothermic [42]. Therefore, from experimental data shown in Tables 6 and 7, it is easy to conclude that CV and CR dyes' adsorption processes in the Pal, Pal-2M, Pal-4M, and Pal-6M samples are predominantly endothermic. The negative ΔG values measured indicated that the adsorption of both dyes in the samples studied was viable and spontaneous thermodynamically [57]. In addition, the positive ΔS values indicated an increased randomness and disorder of the adsorbent surface after adsorption [105].

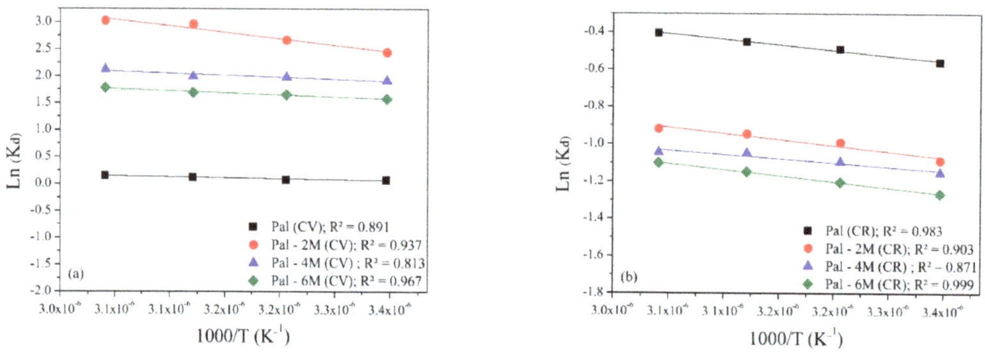

Figure 11. Van't Hoff plot for the removal of (**a**) CV and (**b**) CR dyes.

Table 6. Thermodynamic parameters for the adsorption of CV dye in the palygorskite samples.

Sample	ΔG (kJ mol^{-1})				ΔH (kJ mol^{-1})	ΔS (J k^{-1} mol^{-1})
	298 K	**308 K**	**318 K**	**328 K**		
Pal	−0.17	−0.20	−0.32	−0.41	0.27	0.97
Pal-2M	−6.07	−6.83	−7.82	−8.23	1.98	9.11
Pal-4M	−4.75	−5.01	−5.29	−5.81	0.63	4.02
Pal-6M	−3.91	−4.25	4.47	−4.82	0.62	3.67

Table 7. Thermodynamic parameters for the adsorption of CR dye in the palygorskite samples.

Sample	ΔG (kJ mol^{-1})				ΔH (kJ mol^{-1})	ΔS (J k^{-1} mol^{-1})
	298 K	**308 K**	**318 K**	**328 K**		
Pal	−1.41	−1.56	−1.69	−1.83	0.49	1.11
Pal-2M	−0.82	−0.95	−1.03	−1.09	0.54	0.76
Pal-4M	−0.77	−0.84	−0.92	−0.95	0.37	0.11
Pal-6M	−0.69	−0.77	−0.85	−0.89	0.54	0.54

3.6. Characterization after Adsorption

The FTIR spectra acquired from the Pal and Pal-2M samples before and after the adsorption of CV and CR are shown in Figure 12a,b. After the CV dye's adsorption on the Pal and Pal-2M samples (Figure 12a), new bands were detected at 1588 cm^{-1} and 1366 cm^{-1}. These bands were due to the stretching vibrations C=C and C-N that are characteristic of the CV dye [106,107]. Such bands confirm the removal of the CV dye molecules from the aqueous solution after adsorption by chemisorption [108].

Figure 12b shows that the FTIR spectra measured from the Pal and Pal-2M samples did not change significantly after CR's adsorption. This indicates that there was no breakage or formation of new bonds after adsorption, suggesting the occurrence of physical adsorption (physisorption) in the adsorption of the CR dye [14].

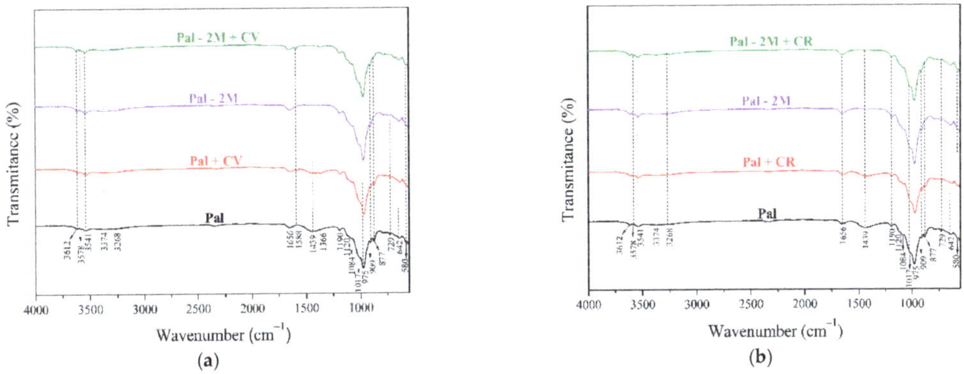

Figure 12. FTIR spectra of the Pal and Pal-2M samples before and after the adsorption of (**a**) CV and (**b**) CR dyes.

The different chemical structures of the dyes result in different forms of bonding on the adsorbent's surface [33]. Figure 13 illustrates the interaction of the CV and CR dye molecules on the palygorskite surface. The positively charged CV molecules are likely to bind almost perpendicular to the clay surface. Conversely, negatively charged CR molecules are likely to bind perpendicularly to the surface. Thus, the molecular structure of the dye also influences the adsorption capacity of the adsorbent.

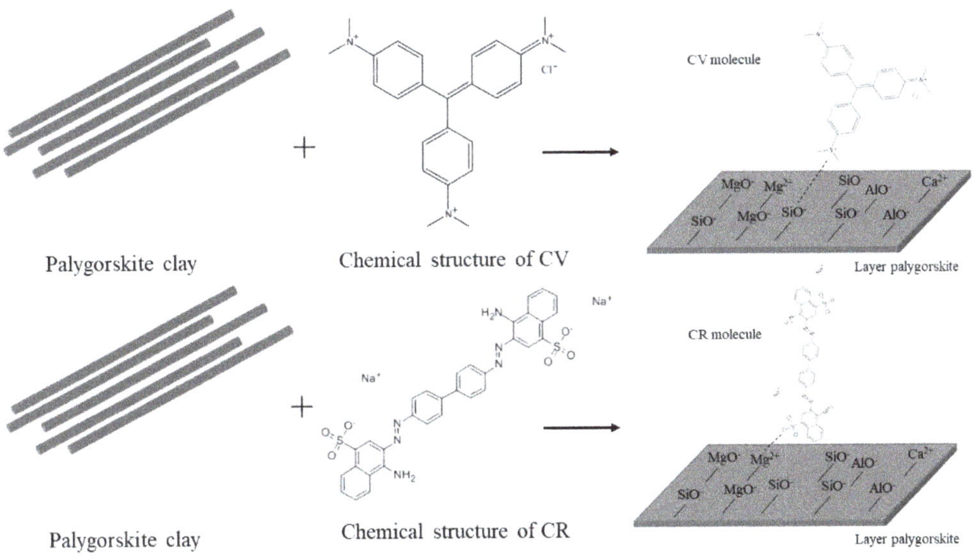

Figure 13. Schema of the interaction between the raw and acid-treated palygorskite surface and CV and CR dyes.

3.7. Regeneration Study

To evaluate the reuse of the Pal and Pal-2M samples, 1 M of CH_3COOH was used to regenerate the adsorbent adsorption sites. After desorption, the regenerated adsorbent was reused for four cycles. The results are shown in Figure 14a,b. Figure 14a shows the effect of regeneration on the adsorption capacity of Pal and Pal-2M samples for CV removal. After four cycles, the decrease in CV removal efficiency was less than 20% for both samples, going from 52% to 30% and from 92% for 76% for Pal and Pal-2M, respectively. Figure 14b shows the effect of regeneration on the adsorption capacity of the Pal and Pal-2M samples for CR removal. After four cycles, the decrease in CR removal efficiency was also less than 20% for both samples, going from 36% to 20% and from 25% for 10% for the Pal and Pal-2M samples, respectively.

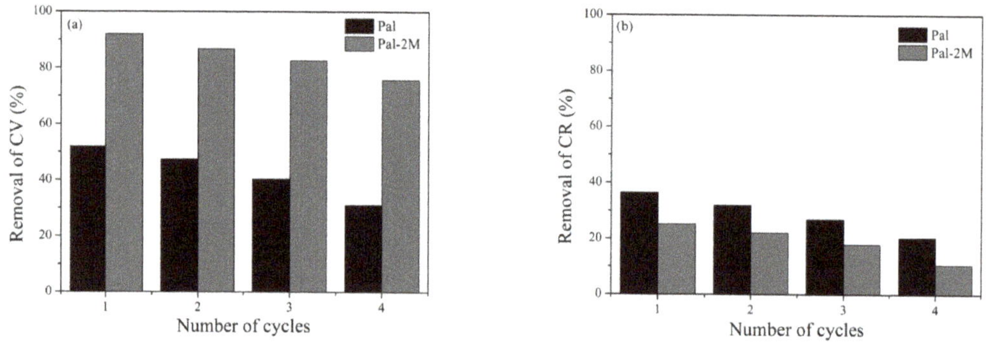

Figure 14. Reusability of the Pal and Pal-2M in four cycles for the removal (**a**) CV and (**b**) CR dyes.

The decrease in the efficiency of Pal and Pal-2M for removing CV and CR dyes occurs due to the dye molecules not being completely released by desorption; some of them may have remained, resulting in repulsive forces for the new dye molecules [99].

3.8. Comparison with Other Adsorbents

Figure 15 compares the maximum amount of CR dye adsorbed in the raw and acid-treated Brazilian palygorskite (Pal, Pal-2M, Pal-4M, and Pal-6M) with works published elsewhere that worked with other adsorbents. For example, the Zn/SiO_2 composite presented a maximum amount of CR adsorbed equal to 83 mg/g [109]; the maximum amount of CR absorbed by chitosan hydrogels was 44.2 mg/g [110], by kaolin it was 5.94 mg/g [111], and by amino-functionalized silica gel it was 10 mg/g [112]. These experimental results indicate that the Brazilian palygorskite has a competitive adsorption capacity for the CR dye.

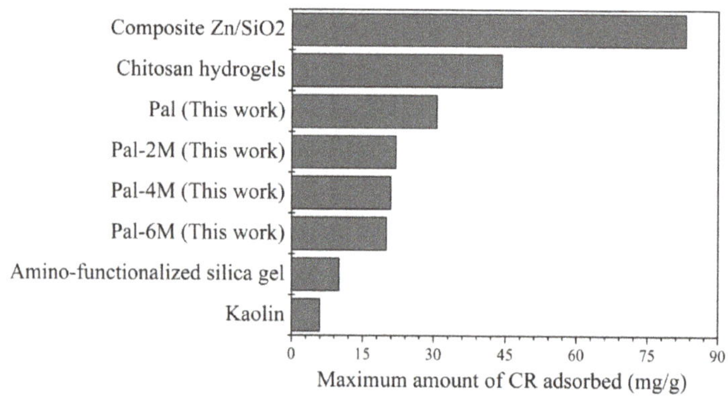

Figure 15. Comparison of the adsorption capacity of different adsorbents for the CR.

Compared with the Pal sample, the Pal-2M, Pal-4M, and Pal-6M samples presented higher CV adsorption capacities. The Pal-2M presented a high adsorption capacity (82.1 mg/g). CV adsorption studies accomplished with others adsorbents presented the following CV adsorption capacities (Figure 16): chitosan-graphite oxide-modified polyurethane (1.98 mg/g) [113], titanate nanotubes (8.36 mg/g) [114], nano mesocellular foam silica (6.64 mg/g) [115], Moroccan clay (17.05 mg/g) [116], magnetically modified activated carbon (67.1 mg/g) [117], Chinese palygorskite treated with HCl [33] (180.24 mg/g) and Oman palygorskite (70 mg/g) [41]. Palygorskites from different regions showed better CV adsorption performances when compared to synthetic adsorbents. The Pal-2M sample showed a lower CV adsorptive capacity when compared to acid-treated Chinese palygorskite [33]. Even so, it presented an adsorption capacity superior to that of other adsorbents found in the literature. This depicts the potential of acid-treated Brazilian palygorskite to be used in the treatment of dye-contaminated wastewater.

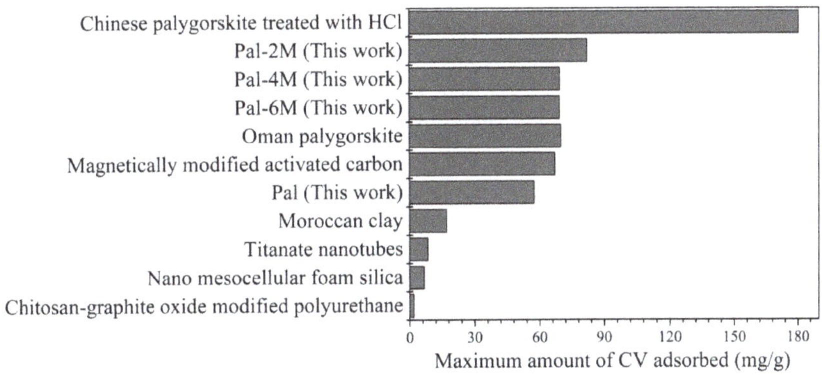

Figure 16. Comparison of the adsorption capacity of different adsorbents for the CV.

Clays are materials with the potential to green the world of materials in the 21st century because of their sustainable character [93]. Thus, the acid-treated palygorskite is a sustainable alternative for the treatment of water contaminated with dyes. The modification technique used in this work is simple, relatively economical, easy to operate, and able to produce adsorption capacities equivalent to those obtained using complex and expensive methods and non-sustainable adsorbent materials. However, only raw palygorskite, as observed in other works, does not seem to be the most efficient adsorption material. The acid-treated palygorskite is an interesting alternative to produce efficient adsorption systems to treat dye-contaminated water.

4. Conclusions

In this study, the crystal violet (CV) adsorption and congo red (CR) dyes on raw and acid-treated Brazilian palygorskite were successfully evaluated. Experiments of chemical analysis, X-ray diffraction, and FTIR showed that the acid treatment used did not destroy the structure of the studied dioctahedral-trioctahedral character palygorskite. Adsorption behavior was not directly associated with BET, and samples treated with the lowest acid concentration solution (2M) showed good efficiency for the cationic dye's adsorption. In basic pH, the acid-treated sample removed almost all the CV dye (97%). In acid pH, the removal of the CR dye was greater than 50%. From a thermodynamics viewpoint, the CV and CR dyes are predominantly endothermic, viable, and spontaneous. The FTIR spectra performed after the CV dye's adsorption on the Pal and Pal-2M samples identified the appearance of new bands at 1588 cm^{-1} and 1366 cm^{-1}, respectively. Such bands confirm the removal of the CV dye molecules from the aqueous solution after adsorption by chemisorption. On the other hand, the FTIR spectra measured from the Pal and Pal-2M samples after adsorption of the CR dye indicated that there was no breakage or formation of new bonds after adsorption, suggesting the occurrence of physical adsorption (physisorption) in the adsorption of the CR dye. The adsorbents derived from Brazilian palygorskite proved to be promising candidates for removing cationic and anionic dyes from water.

Author Contributions: Experimental data, data curation, and formal analysis, V.C.S., M.E.B.A., A.M.R. and R.R.M.; conceptualization, funding acquisition, and project administration, G.A.N., R.R.M. and J.M.C.; formal analysis and writing—review and editing, V.C.S., R.R.M. and A.M.R. All authors have read and agreed to the published version of the manuscript.

Funding: This research was funded by the Coordenação de Aperfeiçoamento de Pessoal e Nível Superior–Brasil (CAPES)–Finance Code 001 (scholarship granted to Vanderlane Cavalcanti da Silva), grant number 88882.455338/2019–01, and National Council for Scientific and Technological Development (CNPq), grant numbers 126892/2020-2, 308822/2018-8, and 420004/2018-1.

Institutional Review Board Statement: Not applicable.

Informed Consent Statement: Not applicable.

Data Availability Statement: Not applicable.

Conflicts of Interest: The authors declare no conflict of interest.

References

1. Goh, P.S.; Ismail, A.F. A review on inorganic membranes for desalination and wastewater treatment. *Desalination* **2018**, *434*, 60–80. [CrossRef]
2. Mirza, A.; Ahmad, R. An efficient sequestration of toxic crystal violet dye from aqueous solution by Alginate/Pectin nanocomposite: A novel and ecofriendly adsorbent. *Groundw. Sustain. Dev.* **2020**, *11*, 100373. [CrossRef]
3. Tyagi, U. Adsorption of dyes using activated carbon derived from pyrolysis of vetiveria zizanioides in a fixed bed reactor. *Groundw. Sustain. Dev.* **2020**, *10*, 100303. [CrossRef]
4. Sirajudheen, P.; Meenakshi, S. Lanthanum (III) incorporated chitosan-montmorillonite composite as flexible material for adsorptive removal of azo dyes from water. *Mater. Today Proc.* **2020**, *27*, 318–326. [CrossRef]
5. Elella, M.H.A.; Sabaa, M.W.; Abd ElHafeez, E.; Mohamed, R.R. Crystal violet dye removal using crosslinked grafted xanthan gum. *Int. J. Biol. Macromol.* **2019**, *137*, 1086–1101. [CrossRef] [PubMed]
6. Ahmad, R.; Ansari, K. Polyacrylamide-Grafted Actinidia deliciosa peels powder (PGADP) for the sequestration of crystal violet dye: Isotherms, kinetics and thermodynamic studies. *Appl. Water Sci.* **2020**, *10*, 1–15. [CrossRef]
7. Cheruiyot, G.K.; Wanyonyi, W.C.; Kiplimo, J.J.; Maina, E.N. Adsorption of toxic crystal violet dye using coffee husks: Equilibrium, kinetics and thermodynamics study. *Sci. Afr.* **2019**, *5*, e00116. [CrossRef]
8. Ebrahimpour, M.; Hassaninejad-Darzi, S.K.; Zavvar Mousavi, H. Adsorption of ternary toxic crystal violet, malachite green and methylene blue onto synthesised SBA-15 mesoporous nanoparticles. *Int. J. Environ. Anal. Chem.* **2020**, 1–24. [CrossRef]
9. Pang, X.; Sellaoui, L.; Franco, D.; Dotto, G.L.; Georgin, J.; Bajahzar, A.; Belmabrouk, H.; Lamine, A.B.; Bonilla-Petriciolet, A.; Li, Z. Adsorption of crystal violet on biomasses from pecan nutshell, para chestnut husk, araucaria bark and palm cactus: Experimental study and theoretical modeling via monolayer and double layer statistical physics models. *Chem. Eng. J.* **2019**, *378*, 122101. [CrossRef]
10. Sadeghi, S.; Nasehi, Z. Simultaneous determination of Brilliant Green and Crystal Violet dyes in fish and water samples with dispersive liquid-liquid micro-extraction using ionic liquid followed by zero crossing first derivative spectrophotometric analysis method. *Spectrochim. Acta Part A Mol. Biomol. Spectrosc.* **2018**, *201*, 134–142. [CrossRef]
11. Ciobanu, G.; Harja, M.; Diaconu, M.; Cimpeanu, C.; Teodorescu, R.; Bucur, D. Crystal violet dye removal from aqueous solution by nanohydroxyapatite. *J. Food Agric. Environ.* **2014**, *12*, 499–502.
12. Mani, S.; Bharagava, R.N. Exposure to crystal violet, its toxic, genotoxic and carcinogenic effects on environment and its degradation and detoxification for environmental safety. *Rev. Environ. Contamin. Toxicol.* **2016**, *237*, 71–104.
13. Kulkarni, M.R.; Revanth, T.; Acharya, A.; Bhat, P. Removal of Crystal Violet dye from aqueous solution using water hyacinth: Equilibrium, kinetics and thermodynamics study. *Resour. Technol.* **2017**, *3*, 71–77. [CrossRef]
14. Li, Z.; Hanafy, H.; Zhang, L.; Sellaoui, L.; Netto, M.S.; Oliveira, M.L.S.; Seliem, M.K.; Dotto, G.L.; Bonilla-Petriciolet, A.; Li, Q. Adsorption of congo red and methylene blue dyes on an ashitaba waste and a walnut shell-based activated carbon from aqueous solutions: Experiments, characterization and physical interpretations. *Chem. Eng. J.* **2020**, *388*, 124263. [CrossRef]
15. Gupta, V.K.; Agarwal, S.; Ahmad, R.; Mirza, A.; Mittal, J. Sequestration of toxic congo red dye from aqueous solution using ecofriendly guar gum/activated carbon nanocomposite. *Int. J. Biol. Macromol.* **2020**, *158*, 1310–1318. [CrossRef]
16. Swan, N.B.; Zaini, M.A.A. Adsorption of malachite green and congo red dyes from water: Recent progress and future outlook. *Ecol. Chem. Eng. S* **2019**, *26*, 119–132. [CrossRef]
17. Elwakeel, K.Z.; Elgarahy, A.M.; Elshoubaky, G.A.; Mohammad, S.H. Microwave assist sorption of crystal violet and Congo red dyes onto amphoteric sorbent based on upcycled Sepia shells. *J. Environ. Health Sci. Eng.* **2020**, *18*, 35–50. [CrossRef]
18. Zhang, Q.; Yan, Z.; Ouyang, J.; Zhang, Y.; Yang, H.; Chen, D. Chemically modified kaolinite nanolayers for the removal of organic pollutants. *Appl. Clay Sci.* **2018**, *157*, 283–290. [CrossRef]
19. Kirankumar, V.S.; Sumathi, S. Copper and cerium co-doped cobalt ferrite nanoparticles: Structural, morphological, optical, magnetic, and photocatalytic properties. *Environ. Sci. Pollut. Res.* **2019**, *26*, 19189–19206. [CrossRef]

20. Cao, X.-L.; Yan, Y.-N.; Zhou, F.-Y.; Sun, S.-P. Tailoring nanofiltration membranes for effective removing dye intermediates in complex dye-wastewater. *J. Membr. Sci.* **2020**, *595*, 117476. [CrossRef]
21. Bouras, H.D.; Isik, Z.; Arikan, E.B.; Bouras, N.; Chergui, A.; Yatmaz, H.C.; Dizge, N. Photocatalytic oxidation of azo dye solutions by impregnation of ZnO on fungi. *Biochem. Eng. J.* **2019**, *146*, 150–159. [CrossRef]
22. Zhou, F.; Cheng, Y.; Gan, L.; Chen, Z.; Megharaj, M.; Naidu, R. Burkholderia vietnamiensis C09V as the functional biomaterial used to remove crystal violet and Cu (II). *Ecotoxicol. Environ. Saf.* **2014**, *105*, 1–6. [CrossRef]
23. Baddouh, A.; El Ibrahimi, B.; Rguitti, M.M.; Mohamed, E.; Hussain, S.; Bazzi, L. Electrochemical removal of methylene bleu dye in aqueous solution using Ti/RuO2–IrO2 and SnO2 electrodes. *Sep. Sci. Technol.* **2019**, 1–10. [CrossRef]
24. Yang, H.-C.; Gong, J.-L.; Zeng, G.-M.; Zhang, P.; Zhang, J.; Liu, H.-Y.; Huan, S.-Y. Polyurethane foam membranes filled with humic acid-chitosan crosslinked gels for selective and simultaneous removal of dyes. *J. Colloid Interface Sci.* **2017**, *505*, 67–78. [CrossRef]
25. Wang, Y.; Su, X.; Xu, Z.; Wen, K.; Zhang, P.; Zhu, J.; He, H. Preparation of surface-functionalized porous clay heterostructures via carbonization of soft-template and their adsorption performance for toluene. *Appl. Surf. Sci.* **2016**, *363*, 113–121. [CrossRef]
26. Sana, D.; Jalila, S. A comparative study of adsorption and regeneration with different agricultural wastes as adsorbents for the removal of methylene blue from aqueous solution. *Chin. J. Chem. Eng.* **2017**, *25*, 1282–1287. [CrossRef]
27. Shi, C.; Tao, F.; Cui, Y. Evaluation of nitriloacetic acid modified cellulose film on adsorption of methylene blue. *Int. J. Biol. Macromol.* **2018**, *114*, 400–407. [CrossRef]
28. Tang, J.; Mu, B.; Zong, L.; Wang, A. From waste hot-pot oil as carbon precursor to development of recyclable attapulgite/carbon composites for wastewater treatment. *J. Environ. Sci.* **2019**, *75*, 346–358. [CrossRef] [PubMed]
29. Jiang, H.-L.; Xu, M.-Y.; Xie, Z.-W.; Hai, W.; Xie, X.-L.; He, F.-A. Selective adsorption of anionic dyes from aqueous solution by a novel β-cyclodextrin-based polymer. *J. Mol. Struct.* **2020**, *1203*, 127373. [CrossRef]
30. Fernandes, J.V.; Rodrigues, A.M.; Menezes, R.R.; Neves, G.D.A. Adsorption of Anionic Dye on the Acid-Functionalized Bentonite. *Materials* **2020**, *13*, 3600. [CrossRef]
31. Buburuzan, A.M.; Catrinescu, C.; Macoveanu, M. Comparative study of the adsorption-desorption cycles of hexane over hypercrosslinked polymeric adsorbents and activated carbon. *Environ. Eng. Manag. J.* **2010**, *9*, 125–132. [CrossRef]
32. Zhu, J.; Zhang, P.; Wang, Y.; Wen, K.; Su, X.; Zhu, R.; He, H.; Xi, Y. Effect of acid activation of palygorskite on their toluene adsorption behaviors. *Appl. Clay Sci.* **2018**, *159*, 60–67. [CrossRef]
33. Yang, R.; Li, D.; Li, A.; Yang, H. Adsorption properties and mechanisms of palygorskite for removal of various ionic dyes from water. *Appl. Clay Sci.* **2018**, *151*, 20–28. [CrossRef]
34. Azha, S.F.; Sellaoui, L.; Yunus, E.H.E.; Yee, C.J.; Bonilla-Petriciolet, A.; Lamine, A.B.; Ismail, S. Iron-modified composite adsorbent coating for azo dye removal and its regeneration by photo-Fenton process: Synthesis, characterization and adsorption mechanism interpretation. *Chem. Eng. J.* **2019**, *361*, 31–40. [CrossRef]
35. Piri, F.; Mollahosseini, A.; Hosseini, M.M. Enhanced adsorption of dyes on microwave-assisted synthesized magnetic zeolite-hydroxyapatite nanocomposite. *J. Environ. Chem. Eng.* **2019**, *7*, 103338. [CrossRef]
36. Xinguo, X.; Jiling, Z.; Ruiyu, J.; Qi, X. Application of Modified Attapulgite Clay as the Adsorbent in Gasoline Desulfurization. *China Pet. Process. Petrochem. Technol.* **2014**, *16*, 63–68.
37. Han, J.; Liang, X.; Xu, Y.; Xu, Y. Removal of Cu2+ from aqueous solution by adsorption onto mercapto functionalized palygorskite. *J. Ind. Eng. Chem.* **2015**, *23*, 307–315. [CrossRef]
38. Zhang, Y.; Wang, W.; Zhang, J.; Liu, P.; Wang, A. A comparative study about adsorption of natural palygorskite for methylene blue. *Chem. Eng. J.* **2015**, *262*, 390–398. [CrossRef]
39. Câmara, A.B.F.; Sales, R.V.; Bertolino, L.C.; Furlanetto, R.P.P.; Rodríguez-Castellón, E.; De Carvalho, L.S. Novel application for palygorskite clay mineral: A kinetic and thermodynamic assessment of diesel fuel desulfurization. *Adsorption* **2020**, *26*, 267–282. [CrossRef]
40. Huang, D.; Zheng, Y.; Zhang, Z.; Quan, Q.; Qiang, X. Synergistic effect of hydrophilic palygorskite and hydrophobic zein particles on the properties of chitosan films. *Mater. Des.* **2020**, *185*, 108229. [CrossRef]
41. Al-Futaisi, A.; Jamrah, A.; Al-Hanai, R. Aspects of cationic dye molecule adsorption to palygorskite. *Desalination* **2007**, *214*, 327–342. [CrossRef]
42. Youcef, L.D.; Belaroui, L.S.; López-Galindo, A. Adsorption of a cationic methylene blue dye on an Algerian palygorskite. *Appl. Clay Sci.* **2019**, *179*, 105145. [CrossRef]
43. Abidi, N.; Duplay, J.; Jada, A.; Errais, E.; Ghazi, M.; Semhi, K.; Trabelsi-Ayadi, M. Removal of anionic dye from textile industries' effluents by using Tunisian clays as adsorbents. Zeta potential and streaming-induced potential measurements. *Comptes Rendus Chim.* **2019**, *22*, 113–125. [CrossRef]
44. Silva, M.P.D.; Santos, M.d.S.F.; Santos, M.R.M.C.; Santos Júnior, L.D.S.; Fonseca, M.G.D.; Silva Filho, E.C.D. Natural palygorskite as an industrial dye remover in single and binary systems. *Mater. Res.* **2016**, *19*, 1232–1240. [CrossRef]
45. Hu, Z.; Shi, F.; Liu, G.; Zhang, B.; Zhang, H. Removal of Methylene Blue from Aqueous Solution by Adsorption Onto Modified Attapulgite Clay. *Energy Environ. Focus* **2015**, *4*, 316–323. [CrossRef]
46. da Costa, F.P.; da Silva Morais, C.R.; Pinto, H.C.; Rodrigues, A.M. Microstructure and physico-mechanical properties of Al2O3-doped sustainable glass-ceramic foams. *Mater. Chem. Phys.* **2020**, *256*, 123612. [CrossRef]
47. Pereira da Costa, F.; Rodrigues da Silva Morais, C.; Rodrigues, A.M. Sustainable glass-ceramic foams manufactured from waste glass bottles and bentonite. *Ceram. Int.* **2020**, *46*, 17957–17961. [CrossRef]

48. Brunauer, S.; Emmett, P.H.; Teller, E. Adsorption of gases in multimolecular layers. *J. Am. Chem. Soc.* **1938**, *60*, 309–319. [CrossRef]
49. Lima, A.; Bruno, C.; Luna, B.; Fl, A.; Alves, S.; Lucena, S.; Margareti, S.; Meneghetti, P. From Disposal to Reuse: Production of Sustainable Fatty Acid Alkyl Esters Derived from Residual Oil Using a Biphasic Magnetic Catalyst. *Sustainability* **2020**, *12*, 10159.
50. Aljeboree, A.M.; Alshirifi, A.N.; Alkaim, A.F. Kinetics and equilibrium study for the adsorption of textile dyes on coconut shell activated carbon. *Arab. J. Chem.* **2017**, *10*, S3381–S3393. [CrossRef]
51. Kuang, Y.; Zhang, X.; Zhou, S. Adsorption of methylene blue in water onto activated carbon by surfactant modification. *Water* **2020**, *12*, 587. [CrossRef]
52. Sahu, O.; Singh, N. Significance of bioadsorption process on textile industry wastewater. In *The Impact and Prospects of Green Chemistry for Textile Technology*; Elsevier: Amsterdam, The Netherlands, 2019; pp. 367–416.
53. Doğan, M.; Özdemir, Y.; Alkan, M. Adsorption kinetics and mechanism of cationic methyl violet and methylene blue dyes onto sepiolite. *Dye. Pigment.* **2007**, *75*, 701–713. [CrossRef]
54. Garba, Z.N. The Relevance of Isotherm and Kinetic Models to Chlorophenols Adsorption: A Review. *Avicenna J. Environ. Health Eng.* **2019**, *6*, 55–65. [CrossRef]
55. Simonin, J.-P. On the comparison of pseudo-first order and pseudo-second order rate laws in the modeling of adsorption kinetics. *Chem. Eng. J.* **2016**, *300*, 254–263. [CrossRef]
56. Staroń, P.; Chwastowski, J.; Banach, M. Sorption and desorption studies on silver ions from aqueous solution by coconut fiber. *J. Clean. Prod.* **2017**, *149*, 290–301. [CrossRef]
57. Mohammadi, N.; Khani, H.; Gupta, V.K.; Amereh, E.; Agarwal, S. Adsorption process of methyl orange dye onto mesoporous carbon material–kinetic and thermodynamic studies. *J. Colloid Interface Sci.* **2011**, *362*, 457–462. [CrossRef]
58. Suarez, M.; García-Romero, E.; Del Río, M.S.; Martinetto, P.; Dooryhee, E. The effect of the octahedral cations on the dimensions of the palygorskite cell. *Clay Miner.* **2007**, *42*, 287–297. [CrossRef]
59. Garciá-Romero, E.; Suárez, M. On the chemical composition of sepiolite and palygorskite. *Clays Clay Miner.* **2010**, *58*, 1–20. [CrossRef]
60. Suárez, M.; García-Rivas, J.; Sánchez-Migallón, J.M.; García-Romero, E. Spanish palygorskites: Geological setting, mineralogical, textural and crystal-chemical characterization. *Eur. J. Mineral.* **2018**, *30*, 733–746. [CrossRef]
61. Wang, Y.; Shen, Y.; Qin, Z.; Li, S.; Zhang, T. Preparation of three-dimensional palygorskite based carrier. *MethodsX* **2020**, *7*, 100815. [CrossRef]
62. Augsburger, M.S.; Strasser, E.; Perino, E.; Mercader, R.C.; Pedregosa, J.C. FTIR and Mössbauer investigation of a substituted palygorskite: Silicate with a channel structure. *J. Phys. Chem. Solids* **1998**, *59*, 175–180. [CrossRef]
63. Chen, H.; Zhao, Y.; Wang, A. Removal of Cu (II) from aqueous solution by adsorption onto acid-activated palygorskite. *J. Hazard. Mater.* **2007**, *149*, 346–354. [CrossRef]
64. Boudriche, L.; Calvet, R.; Hamdi, B.; Balard, H. Effect of acid treatment on surface properties evolution of attapulgite clay: An application of inverse gas chromatography. *Colloids Surf. A Physicochem. Eng. Asp.* **2011**, *392*, 45–54. [CrossRef]
65. Moulton, B.J.A.; Rodrigues, A.M.; Sampaio, D.V.; Silva, L.D.; Cunha, T.R.; Zanotto, E.D.; Pizani, P.S. The origin of the unusual DSC peaks of supercooled barium disilicate liquid. *CrystEngComm* **2019**, *21*, 2768–2778. [CrossRef]
66. Moulton, B.J.A.; Rodrigues, A.M.; Pizani, P.S.; Sampaio, D.V.; Zanotto, E.D. A Raman investigation of the structural evolution of supercooled liquid barium disilicate during crystallization. *Int. J. Appl. Glas. Sci.* **2018**, *9*, 510–517. [CrossRef]
67. Marosz, M.; Kowalczyk, A.; Gil, B.; Chmielarz, L. Acid-treated Clay Minerals as Catalysts for Dehydration of Methanol and Ethanol. *Clays Clay Miner.* **2020**, *68*, 23–37. [CrossRef]
68. Xavier, K.C.M.; Santos, M.D.S.F.D.; Santos, M.R.M.C.; Oliveira, M.E.R.; Carvalho, M.W.N.C.; Osajima, J.A.; Silva Filho, E.C.D. Effects of acid treatment on the clay palygorskite: XRD, surface area, morphological and chemical composition. *Mater. Res.* **2014**, *17*, 3–8. [CrossRef]
69. Oliveira, R.N.; Acchar, W.; Soares, G.D.A.; Barreto, L.S. The increase of surface area of a Brazilian palygorskite clay activated with sulfuric acid solutions using a factorial design. *Mater. Res.* **2013**, *16*, 924–928. [CrossRef]
70. Dhanya, B.A.P.; Pushpaletha, P. Metal supported and metal ion exchanged catalysts from palygorskite for acetylation reaction. *Indian J. Chem. A* **2020**, *57*, 649–654.
71. Madejova, J.; Komadel, P. Baseline studies of the clay minerals society source clays: Infrared methods. *Clays Clay Miner.* **2001**, *49*, 410–432. [CrossRef]
72. Suarez, M.; Garcia-Romero, E. FTIR spectroscopic study of palygorskite: Influence of the composition of the octahedral sheet. *Appl. Clay Sci.* **2006**, *31*, 154–163. [CrossRef]
73. Tian, G.; Wang, W.; Kang, Y.; Wang, A. Palygorskite in sodium sulphide solution via hydrothermal process for enhanced methylene blue adsorption. *J. Taiwan Inst. Chem. Eng.* **2016**, *58*, 417–423. [CrossRef]
74. Frost, R.L.; Locos, O.B.; Ruan, H.; Kloprogge, J.T. Near-infrared and mid-infrared spectroscopic study of sepiolites and palygorskites. *Vib. Spectrosc.* **2001**, *27*, 1–13. [CrossRef]
75. Frost, R.L.; Bahfenne, S.; Graham, J. Infrared and infrared emission spectroscopic study of selected magnesium carbonate minerals containing ferric iron—Implications for the geosequestration of greenhouse gases. *Spectrochim. Acta Part A Mol. Biomol. Spectrosc.* **2008**, *71*, 1610–1616. [CrossRef]
76. Liu, Q.; Yao, X.; Cheng, H.; Frost, R.L. An infrared spectroscopic comparison of four Chinese palygorskites. *Spectrochim. Acta Part A Mol. Biomol. Spectrosc.* **2012**, *96*, 784–789. [CrossRef]

77. Rhouta, B.; Zatile, E.; Bouna, L.; Lakbita, O.; Maury, F.; Daoudi, L.; Lafont, M.C.; Amjoud, M.; Senocq, F.; Jada, A. Comprehensive physicochemical study of dioctahedral palygorskite-rich clay from Marrakech High Atlas (Morocco). *Phys. Chem. Miner.* **2013**, *40*, 411–424. [CrossRef]
78. Tian, G.; Wang, W.; Mu, B.; Kang, Y.; Wang, A. Facile fabrication of carbon/attapulgite composite for bleaching of palm oil. *J. Taiwan Inst. Chem. Eng.* **2015**, *50*, 252–258. [CrossRef]
79. Ouyang, J.; Zhao, Z.; Suib, S.L.; Yang, H. Degradation of Congo Red dye by a Fe2O3@ CeO2-ZrO2/Palygorskite composite catalyst: Synergetic effects of Fe2O3. *J. Colloid Interface Sci.* **2019**, *539*, 135–145. [CrossRef]
80. Zhang, J.; Wang, Q.; Chen, H.; Wang, A. XRF and nitrogen adsorption studies of acid-activated palygorskite. *Clay Miner.* **2010**, *45*, 145–156. [CrossRef]
81. Rusmin, R.; Sarkar, B.; Biswas, B.; Churchman, J.; Liu, Y.; Naidu, R. Structural, electrokinetic and surface properties of activated palygorskite for environmental application. *Appl. Clay Sci.* **2016**, *134*, 95–102. [CrossRef]
82. Frinisrasra, N.; Srasra, E. Effect of heating on palygorskite and acid treated palygorskite properties. *Электронная обработка материалов* **2008**, *44*, 43–49.
83. Xavier, K.C.M.; Santos, M.S.F.; Osajima, J.A.; Luz, A.B.; Fonseca, M.G.; Silva Filho, E.C. Thermally activated palygorskites as agents to clarify soybean oil. *Appl. Clay Sci.* **2016**, *119*, 338–347. [CrossRef]
84. Ogorodova, L.; Vigasina, M.; Melchakova, L.; Krupskaya, V.; Kiseleva, I. Thermochemical study of natural magnesium aluminum phyllosilicate: Palygorskite. *J. Chem. Thermodyn.* **2015**, *89*, 205–211. [CrossRef]
85. Bayram, H.; Önal, M.; Üstünışık, G.; Sarıkaya, Y. Some thermal characteristics of a mineral mixture of palygorskite, metahalloysite, magnesite and dolomite. *J. Therm. Anal. Calorim.* **2007**, *89*, 169–174. [CrossRef]
86. Amorim, K.B.; Angélica, R.S. Mineralogy and geochemistry of occurrence of palygorskite of Alcântara, S. Luís-Grajaú basin, Maranhão, Brazil. *Cerâmica* **2011**, *57*, 483–490. [CrossRef]
87. Wang, K.; Wang, L.; Zhang, Y.; Zhang, Y.; Liang, J. Microstructural evolution and sintering properties of palygorskite nanofibers. *Int. J. Appl. Ceram. Technol.* **2020**, *17*, 1833–1842. [CrossRef]
88. Bouna, L.; Rhouta, B.; Amjoud, M.; Maury, F.; Lafont, M.-C.; Jada, A.; Senocq, F.; Daoudi, L. Synthesis, characterization and photocatalytic activity of TiO2 supported natural palygorskite microfibers. *Appl. Clay Sci.* **2011**, *52*, 301–311. [CrossRef]
89. Kadïr, S.; Eren, M.; Atabey, E. Dolocretes and associated palygorskite occurrences in siliciclastic red mudstones of the Sariyer formation (Middle Miocene), southeastern side of the Çanakkale strait, Turkey. *Clays Clay Miner.* **2010**, *58*, 205–219. [CrossRef]
90. Myriam, M.; Suarez, M.; Martin-Pozas, J.M. Structural and textural modifications of palygorskite and sepiolite under acid treatment. *Clays Clay Miner.* **1998**, *46*, 225–231. [CrossRef]
91. Gonzalez, F.; Pesquera, C.; Benito, I.; Mendioroz, S.; Pajares, J.A. Mechanism of acid activation of magnesic palygorskite. *Clays Clay Miner.* **1989**, *37*, 258–262. [CrossRef]
92. Omer, O.S.; Hussein, B.H.M.; Ouf, A.M.; Hussein, M.A.; Mgaidi, A. An organified mixture of illite-kaolinite for the removal of Congo red from wastewater. *J. Taibah Univ. Sci.* **2018**, *12*, 858–866. [CrossRef]
93. Wang, W.; Tian, G.; Zhang, Z.; Wang, A. A simple hydrothermal approach to modify palygorskite for high-efficient adsorption of methylene blue and Cu (II) ions. *Chem. Eng. J.* **2015**, *265*, 228–238. [CrossRef]
94. Dong, W.; Lu, Y.; Wang, W.; Zong, L.; Zhu, Y.; Kang, Y.; Wang, A. A new route to fabricate high-efficient porous silicate adsorbents by simultaneous inorganic-organic functionalization of low-grade palygorskite clay for removal of Congo red. *Microporous Mesoporous Mater.* **2019**, *277*, 267–276. [CrossRef]
95. Sousa, H.R.; Silva, L.S.; Sousa, P.A.A.; Sousa, R.R.M.; Fonseca, M.G.; Osajima, J.A.; Silva-Filho, E.C. Evaluation of methylene blue removal by plasma activated palygorskites. *J. Mater. Res. Technol.* **2019**, *8*, 5432–5442. [CrossRef]
96. Zhang, S.; Zhong, L.; Yang, H.; Tang, A.; Zuo, X. Magnetic carbon-coated palygorskite loaded with cobalt nanoparticles for Congo Red removal from waters. *Appl. Clay Sci.* **2020**, *198*, 105856. [CrossRef]
97. Wong, S.; Abd Ghafar, N.; Ngadi, N.; Razmi, F.A.; Inuwa, I.M.; Mat, R.; Amin, N.A.S. Effective removal of anionic textile dyes using adsorbent synthesized from coffee waste. *Sci. Rep.* **2020**, *10*, 1–13. [CrossRef]
98. Wekoye, J.N.; Wanyonyi, W.C.; Wangila, P.T.; Tonui, M.K. Kinetic and equilibrium studies of Congo red dye adsorption on cabbage waste powder. *Environ. Chem. Ecotoxicol.* **2020**, *2*, 24–31. [CrossRef]
99. Puri, C.; Sumana, G. Highly effective adsorption of crystal violet dye from contaminated water using graphene oxide intercalated montmorillonite nanocomposite. *Appl. Clay Sci.* **2018**, *166*, 102–112. [CrossRef]
100. Tang, H.; Li, W.; Zhang, T.; Li, Q.; Xing, J.; Liu, H. Improvement in diesel desulfurization capacity by equilibrium isotherms analysis. *Sep. Purif. Technol.* **2011**, *78*, 352–356. [CrossRef]
101. Litefti, K.; Freire, M.S.; Stitou, M.; González-Álvarez, J. Adsorption of an anionic dye (Congo red) from aqueous solutions by pine bark. *Sci. Rep.* **2019**, *9*, 1–11. [CrossRef]
102. Kaur, S.; Rani, S.; Mahajan, R.K. Adsorption kinetics for the removal of hazardous dye congo red by biowaste materials as adsorbents. *J. Chem.* **2012**. [CrossRef]
103. Khan, M.A.; Uddin, M.K.; Bushra, R.; Ahmad, A.; Nabi, S.A. Synthesis and characterization of polyaniline Zr (IV) molybdophosphate for the adsorption of phenol from aqueous solution. *React. Kinet. Mech. Catal.* **2014**, *113*, 499–517. [CrossRef]
104. Zhu, Y.; Chen, T.; Liu, H.; Xu, B.; Xie, J. Kinetics and thermodynamics of Eu (III) and U (VI) adsorption onto palygorskite. *J. Mol. Liq.* **2016**, *219*, 272–278. [CrossRef]

105. Sahu, S.; Pahi, S.; Tripathy, S.; Singh, S.K.; Behera, A.; Sahu, U.K.; Patel, R.K. Adsorption of methylene blue on chemically modified lychee seed biochar: Dynamic, equilibrium, and thermodynamic study. *J. Mol. Liq.* **2020**, *315*, 113743. [CrossRef]
106. Abdi, M.; Balagabri, M.; Karimi, H.; Hossini, H.; Rastegar, S.O. Degradation of crystal violet (CV) from aqueous solutions using ozone, peroxone, electroperoxone, and electrolysis processes: A comparison study. *Appl. Water Sci.* **2020**, *10*, 1–10. [CrossRef]
107. Cheriaa, J.; Khaireddine, M.; Rouabhia, M.; Bakhrouf, A. Removal of triphenylmethane dyes by bacterial consortium. *Sci. World J.* **2012**. [CrossRef]
108. Jabar, J.M.; Odusote, Y.A.; Alabi, K.A.; Ahmed, I.B. Kinetics and mechanisms of congo-red dye removal from aqueous solution using activated Moringa oleifera seed coat as adsorbent. *Appl. Water Sci.* **2020**, *10*, 1–11. [CrossRef]
109. Zhang, J.; Yan, X.; Hu, M.; Hu, X.; Zhou, M. Adsorption of Congo red from aqueous solution using ZnO-modified SiO2 nanospheres with rough surfaces. *J. Mol. Liq.* **2018**, *249*, 772–778. [CrossRef]
110. El-Harby, N.F.; Ibrahim, S.M.A.; Mohamed, N.A. Adsorption of Congo red dye onto antimicrobial terephthaloyl thiourea cross-linked chitosan hydrogels. *Water Sci. Technol.* **2017**, *76*, 2719–2732. [CrossRef]
111. Zenasni, M.; Meroufel, B.; Merlin, A.; George, B. Adsorption of Congo red from aqueous solution using CTAB-kaolin from Bechar Algeria. *J. Surf. Eng. Mater. Adv. Technol.* **2014**, *4*, 332–341. [CrossRef]
112. Farias, R.S.D.; Buarque, H.L.D.B.; Cruz, M.R.D.; Cardoso, L.M.F.; Gondim, T.D.A.; Paulo, V.R.D. Adsorption of congo red dye from aqueous solution onto amino-functionalized silica gel. *Eng. Sanit. e Ambient.* **2018**, *23*, 1053–1060. [CrossRef]
113. Qin, J.; Qiu, F.; Rong, X.; Yan, J.; Zhao, H.; Yang, D. Adsorption behavior of crystal violet from aqueous solutions with chitosan–graphite oxide modified polyurethane as an adsorbent. *J. Appl. Polym. Sci.* **2015**, *132*, 41828. [CrossRef]
114. Mohanty, S.; Moulick, S.; Maji, S.K. Adsorption/photodegradation of crystal violet (basic dye) from aqueous solution by hydrothermally synthesized titanate nanotube (TNT). *J. Water Process Eng.* **2020**, *37*, 101428. [CrossRef]
115. Zhai, Q.-Z. Studies of adsorption of crystal violet from aqueous solution by nano mesocellular foam silica: Process equilibrium, kinetic, isotherm, and thermodynamic studies. *Water Sci. Technol.* **2020**, *81*, 2092–2108. [CrossRef] [PubMed]
116. Miyah, Y.; Lahrichi, A.; Idrissi, M.; Anis, K.; Kachkoul, R.; Idrissi, N.; Lairini, S.; Nenov, V.; Zerrouq, F. Removal of cationic dye "crystal violet" in aqueous solution by the local clay. *J. Mater. Environ. Sci.* **2017**, *8*, 3570–3582.
117. Hamidzadeh, S.; Torabbeigi, M.; Shahtaheri, S.J. Removal of crystal violet from water by magnetically modified activated carbon and nanomagnetic iron oxide. *J. Environ. Health Sci. Eng.* **2015**, *13*, 8. [CrossRef]

sustainability

MDPI

Article

Production of Eco-Sustainable Materials: Compatibilizing Action in Poly (Lactic Acid)/High-Density Biopolyethylene Bioblends

Eduardo da Silva Barbosa Ferreira [1,*], Carlos Bruno Barreto Luna [1], Danilo Diniz Siqueira [1], Edson Antonio dos Santos Filho [1], Edcleide Maria Araújo [1] and Renate Maria Ramos Wellen [2]

1 Department of Materials Engineering, Federal University of Campina Grande, Campina Grande 58429-900, Brazil; brunobarretodemaufcg@hotmail.com (C.B.B.L.); danilodinizsiqueira@gmail.com (D.D.S.); edson.a.santos.f@gmail.com (E.A.d.S.F.); edcleide.araujo@ufcg.edu.br (E.M.A.)

2 Department of Materials Engineering, Federal University of Paraíba, João Pessoa 58051-085, Brazil; wellen.renate@gmail.com

* Correspondence: eduardosbf95@gmail.com

Abstract: Motivated by environment preservation, the increased use of eco-friendly materials such as biodegradable polymers and biopolymers has raised the interest of researchers and the polymer industry. In this approach, this work aimed to produce bioblends using poly (lactic acid) (PLA) and high-density biopolyethylene (BioPE); due to the low compatibility between these polymers, this work evaluated the additional influence of the compatibilizing agents: poly (ethylene octene) and ethylene elastomer grafted with glycidyl methacrylate (POE-g-GMA and EE-g-GMA, respectively), polyethylene grafted with maleic anhydride (PE-g-MA), polyethylene grafted with acrylic acid (PE-g-AA) and the block copolymer styrene (ethylene-butylene)-styrene grafted with maleic anhydride (SEBS-g-MA) to the thermal, mechanical, thermomechanical, wettability and morphological properties of PLA/BioPE. Upon the compatibilizing agents' addition, there was an increase in the degree of crystallinity observed by DSC (2.3–7.6% related to PLA), in the thermal stability as verified by TG (6–15 °C for $T_{D10\%}$, 6–11 °C $T_{D50\%}$ and 112–121 °C for $T_{D99.9\%}$ compared to PLA) and in the mechanical properties such as elongation at break (with more expressive values for the addition of POE-g-GMA and SEBS-g-MA, 9 and 10%, respectively), tensile strength (6–19% increase compared to PLA/BioPE bioblend) and a significant increase in impact strength, with evidence of plastic deformation as observed through SEM, promoted by the PLA/ BioPE phases improvement. Based on the gathered data, the added compatibilizers provided higher performing PLA/BioPE. The POE-g-GMA compatibilizer was considered to provide the best properties in relation to the PLA/BioPE bioblend, as well as the PLA matrix, mainly in relation to impact strength, with an increase of approximately 133 and 100% in relation to PLA and PLA/BioPE bioblend, respectively. Therefore, new ecological materials can be manufactured, aiming at benefits for the environment and society, contributing to sustainable development and stimulating the consumption of eco-products.

Keywords: poly (lactic acid); high-density biopolyethylene; bioblends; compatibilization

Citation: Ferreira, E.d.S.B.; Luna, C.B.B.; Siqueira, D.D.; dos Santos Filho, E.A.; Araújo, E.M.; Wellen, R.M.R. Production of Eco-Sustainable Materials: Compatibilizing Action in Poly (Lactic Acid)/High-Density Biopolyethylene Bioblends. *Sustainability* **2021**, *13*, 12157. https://doi.org/10.3390/su132112157

Academic Editor: Maria L. Auad

Received: 3 September 2021
Accepted: 12 October 2021
Published: 4 November 2021

1. Introduction

Motivated by the great research growth of biodegradable and polymers obtained from renewable sources (biopolymers), the study and production of bioblends have gained attention from both industry and polymer scientists, due to its eco-friendly character, as well as the search for new properties, providing new systems with improved performance [1–3]. However, due to the interaction deficiency in the polymer blends, immiscible mixtures are often obtained, exhibiting coarse morphology poorly distributed in the matrix, as also low interfacial adhesion between the phases. Therefore, compatibilization is necessary to modify the interfacial properties of these mixtures, leading to interfacial tension reduction and lower coalescence levels [4–11].

Biodegradable and biopolymers have found a rightful place in the polymer industry due to increased environmental attention, as well as the vast versatility to be used in a range of applications such as biomedical items, packaging and general goods for instance [12,13]. Due to its ability to replace material obtained from non-renewable sources, with high elastic modulus and tensile strength, PLA is the thermoplastic aliphatic polyester among those most studied biodegradable polymers in the last 20 years. Nevertheless, limitations are verified, including low elongation at break and impact strength, low toughness which limit its use in some applications [12,14–17].

Among the biopolymers, biopolyethylene (BioPE), produced by the Brazilian petrochemical Braskem, since 2010, has gained prominence for being considered a technological innovation, due to the reduced dependence on fossil materials, as it comes from sugar cane, as also by the CO_2 absorption from the atmosphere during the production cycle, being chemically and displaying equivalent properties to petroleum-based polyethylene [18–20].

Literature reports works based on the production and characterization of PLA/PE blends [21–27]. Ferri et al. [27] studied PLA/BioPE (80/20) bioblends compatibilized with ethylene vinyl acetate (EVA), polyvinyl alcohol (PVA) and dicumyl peroxide (DCP). Due to immiscibility, the binary blend has reduced mechanical properties compared to PLA, however, upon the addition of three compatibilizers (EVA, PVA and DCP), interactions were improved, and as a consequence, properties increase, such as elongation at break and impact strength. Nevertheless, works based on poly (lactic acid)/high-density polyethylene bioblends from sugarcane together with compatibilizers addition such as POE-g-GMA, EE-g-GMA and SEBS-g-MA are still scarce in the specialized literature, making this research topic pertinent.

With a more demanding society for new green technologies, sustainable consumption is increased by ecological materials. It is a practice related to the acquisition of eco-products that aim to minimize impacts on the environment and, at the same time, maintaining the ecological balance on our planet. In recent years, environmentally responsible practices have become part of the strategy of large companies in the field of polymer technology. In view of the positive contribution to sustainability, bioblends are being developed, aiming at a more sustainable environmental cycle. These are manufactured based on materials from proper environmental sources, with clean production technologies and using renewable sources. Therefore, the environmental, social and economic aspects are justifications for researching the production of these ecological materials.

Based on the above mentioned, the objective of this work was producing PLA/BioPE bioblends compatibilized with several agents, i.e., poly (ethylene octene) grafted with glycidyl methacrylate (POE-g-GMA), ethylene elastomer grafted with glycidyl methacrylate (EE-g-GMA), polyethylene grafted with maleic anhydride (PE-g-MA), polyethylene grafted with acrylic acid (PE-g-AA) and the block copolymer styrene-(ethylene-butylene)-styrene grafted with maleic anhydride (SEBS-g-MA), and evaluating the effect of each compatibilizer in the thermal, mechanical, thermomechanical, wettability and morphology properties of produced bioblends.

2. Methodology

2.1. Materials

Poly (lactic acid) from NatureWorks as pellets with a density of 1.24 g/cm^3. High-Density Polyethylene I'm green SHC7260, from Braskem as pellets with density 0.959 g/cm^3, and melting flow rate (MFR) 7.2 g/10 min (190 °C/2.16 kg). As compatibilizing agents: Poly (ethylene octene) grafted with 0.8% glycidyl methacrylate (POE-g-GMA) with trade name Coace W5B from Xiamen Coace Plastic Technology, with density 0.91 g/cm^3 and 3 < MFR < 8 g/10 min; Ethylene elastomer grafted with 0.8% glycidyl methacrylate (EE-g-GMA), trade name Coace W5D, with density 0.92 g/cm^3 and 8 < MFR < 16 g/10 min from Xiamen Coace Plastic Technology; Polyethylene grafted with 1.5–1.7% Maleic Anhydride (PE-g-MA) marketed as Polybond 3029, MFR 4 g/10 min supplied by Addivant; Polyethylene grafted with 5.5–6.5% Acrylic Acid (PE-g-AA) marketed as Polybond 1009,

MFR 5 g/10 min supplied by Addivant, and styrene(ethylene-butylene)-styrene functionalized block copolymer with 1.7% maleic anhydride (SEBS-g-MA) marketed as Kraton® FG1901G with MFR 5 g/10 min (200 °C/5 kg), supplied by the Kraton Polymers Group of Companies.

2.2. Bioblends Processing

Processed bioblends with compositions and codes are shown in Table 1. Initially, bioblends components were manually cold-mixed in order to promote greater homogenization, then extruded using a co-rotational, interpenetrating, modular ZSK model twin-screw extruder (D = 18 mm and L/D = 40), Werner-Pfleiderer, Coperion (Stuttgart, Germany). The processing parameters are shown in Table 2. The screw profile used was configured with distributive and dispersive modules, for better mixture homogeneity.

Table 1. Compositions of binary and compatible bioblends with mass proportion (%).

Samples	PLA (%)	BioPE (%)	POE-g-GMA (%)	EE-g-GMA (%)	PE-g-MA (%)	PE-g-AA (%)	SEBS-g-MA (%)
PLA	100	-	-	-	-	-	-
BioPE	-	100	-	-	-	-	-
PLA/BioPE	70	30	-	-	-	-	-
PLA/BioPE/POE-g-GMA	70	20	10	-	-	-	-
PLA/BioPE/EE-g-GMA	70	20	-	10	-	-	-
PLA/BioPE/PE-g-MA	70	20	-	-	10	-	-
PLA/BioPE/PE-g-AA	70	20	-	-	-	10	-
PLA/BioPE/SEBS-g-MA	70	20	-	-	-	-	10

Table 2. Applied parameters during extrusion and injection processing.

Parameters			
Extrusion		**Injection**	
Feed rate (kg/h)	3	**Injection and hold pressing (bar)**	800 and 500
Temperature profile (°C)	170, 170, 175, 175, 175, 180, 180	**Temperature profile (°C)**	170, 175, 175, 180, 180
Screw rate (rpm)	250	**Mold temperature (°C)**	20

The extruded was pelletized and vacuum dried at 60 °C for 24 h. Afterwards, specimens were injection molded using an Arburg Model Allrounder 207C Golden Edition injection molding machine (Radevormwald, Germany). Tensile, impact, and heat deflection temperature (HDT) specimens were molded according to ASTM D638 (Type I), ASTM D256, and ASTM D648, respectively. Table 2 presents applied parameters during extrusion and specimen injection.

2.3. Characterizations

Differential scanning calorimetry (DSC) analyses were carried out using a TA Instruments DSC-Q20 (New Castle, United States), and samples were heated from room temperature (~23 °C) to 200 °C, at 10 °C/min, under a nitrogen atmosphere with gas flow 50 mL/min. Tested samples were approximately 5 mg weight. The degree of crystallinity (X_c) of produced samples was calculated according to Equation (1) [28]:

$$X_c(\%) = \frac{\Delta H_m - \Delta H_{cc}}{W \times \Delta H_m^o} \times 100 \quad (1)$$

where: ΔH_m is the melting enthalpy of PLA and BioPE; ΔH_{cc} is the cold crystallization enthalpy; ΔH^o is equilibrium melting enthalpy, where 100% crystalline PLA is 93.7 J/g [29], and 100% crystalline BioPE is 290 J/g [30]; W is the matrix content.

Plots of molten fraction and melting rate were acquired through DSC peaks integration using INTEGRAL software, and Equations (2)–(4). The molten fraction (x) as time function (t) is given by the area between (J) and a virtual baseline during the event (J_0), defined as the straight line between the start and end points, where E_0 is the total latent heat of the phase change; and t_1 and t_2 are the start and end times of the event, respectively [31–33]:

$$x(t) = \frac{1}{E_0}\int_{t_1}^{t} |J(t') - J_0(t')|dt' \tag{2}$$

where:

$$E_0 = \int_{t_1}^{t_2} |J(t) - J_0(t)|dt \tag{3}$$

Molten fraction can be expressed as temperature function (T), knowing the linear relationship between time and temperature during the event: $T = T_1 + \varphi\ (t - t_1)$, where T_1 is the sample temperature at the initial point t_1, $\tau = t - t_1$ is the time measured since the event start, and $\varphi = dT/dt$ is the heating/cooling rate (constant) during the event. Therefore, the melting rate (C_m) can be given by Equation (4), from which the melting rate was computed [31–33]:

$$c_m = \frac{dx}{dt} = \frac{|J(t) - J_0(t)|}{E_0} \tag{4}$$

Thermogravimetry (TG) analyses carried out in a TA Instruments SDT Q600 simultaneous TG/DSC device (Kyoto, Japan) employing samples with 5 mg, heated from room temperature (~23 °C) to 500 °C, the heating rate was 10 °C/min and nitrogen flow rate of 100 mL/min.

Tensile test was performed on injected specimens according to ASTM D638 using an EMIC DL 2000 (São José dos Pinhais, Brazil) universal testing machine with an elongation rate of 50 mm/min and load cell of 20 kN at room temperature (~23 °C). Presented results are an average of ten specimens.

Izod impact strength measurements were performed based on ASTM D256-97 using notched specimens in a Ceast Resil 5.5 J device (Turin, Italy) operating with 2.75 J hammer at room temperature (~23 °C). Presented results are an average of ten specimens.

Heat deflection temperature (HDT) was evaluated according to ASTM D648, in a Ceast model HDT 6 VICAT (Turin, Italy) with a voltage of 1.82 MPa, and heating rate 120 °C/h (method A). HDT was determined after specimen deflecting 0.25 mm. Presented results are an average of three specimens.

Vicat softening temperature (VST) test carried out according to the ASTM D1525 standard, in Ceast equipment (Turin, Italy), model HDT 6 VICAT/N 6921.000, at a heating rate of 120 °C/h. VST was determined after the needle penetrating 1 mm into the specimens. Reported results are an average of three experiments.

Shore D hardness test carried out according to the ASTM D2240 standard, on Shore-Durometer Hardness Type "D" Woltest (São Paulo, Brazil), with 50 N load controlled by calibrated springs using standardized indenters. Reported results are an average of five experiments.

The contact angle analysis was performed using the sessile drop method, using a portable contact angle; model Phoenix-i from Surface Electro Optics–SEO (Gyeonggi-do, South Korea). The drop was deposited on impact specimens using a micrometric meter; the image was captured and analyzed through the equipment software.

Scanning electron microscopy (SEM) images were captured on the fractured surface from the impact test. A scanning electron microscope, VEGA 3 TESCAN (Brno, Czech Republic), at a voltage of 30 kV under a high vacuum was used. Fractured surfaces were gold coated.

3. Results and Discussion

3.1. Differential Scanning Calorimetry (DSC)

DSC scans acquired during cooling and second heating for PLA, BioPE, PLA/BioPE and compatibilized bioblends are displayed in Figure 1 and the computed parameters are informed in Table 3.

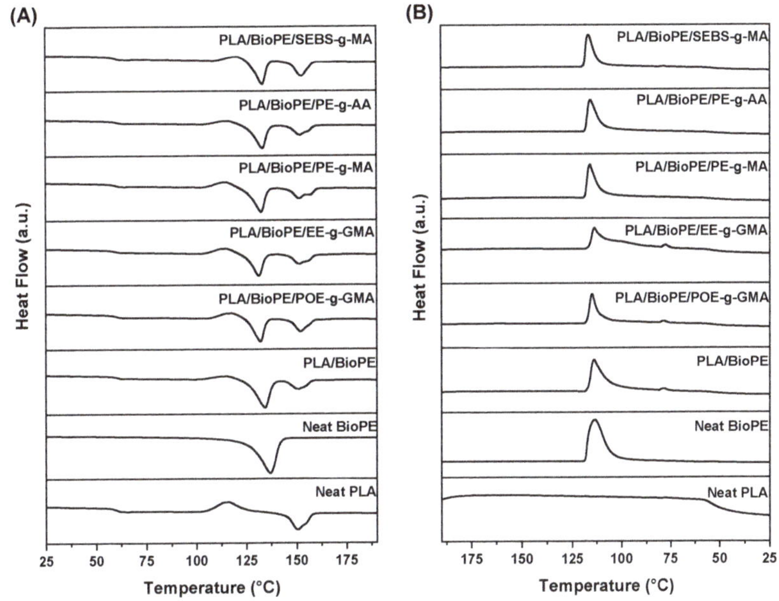

Figure 1. DSC scans acquired during the second heating (**A**) and cooling (**B**) of PLA, BioPE, PLA/BioPE and compatibilized bioblends.

Table 3. Melting and crystallization parameters of investigated samples.

Samples	T_g (°C)	$T_m{}^1$ (°C)	$T_m{}^2$ (°C)	$\Delta H_m{}^1$ (J/g)	$\Delta H_m{}^2$ (J/g)	X_c (%)	T_{cc} (°C)	ΔH_{cc} (J/g)	T_c (°C)
PLA	61	150.4	-	24.5	-	3.9	115.4	20.8	-
BioPE	-	-	136.8	-	176.1	60.1	-	-	113.6
PLA/BioPE	61	150.7	134.3	13.4	39.7	9.4	114.8	7.2	113.9
PLA/BioPE/POE-g-GMA	60	152.7	132.1	16.5	26.1	9.3	117.3	10.4	115
PLA/BioPE/EE-g-GMA	59	151.4	131.3	16.1	27.1	6.2	114.7	12	113.8
PLA/BioPE/PE-g-MA	60	151.6	132.5	19	32.8	11.4	114.7	11.5	116.1
PLA/BioPE/PE-g-AA	61	151.9	133.2	15.4	32.7	10.9	115.8	8.3	115.9
PLA/BioPE/SEBS-g-MA	60	152.7	133.1	16.5	23.1	11.5	120.4	8.9	116.8

T_g is the glass transition temperature; $T_m{}^1$ PLA melting peak temperature; $T_m{}^2$ BioPE melting peak temperature; $\Delta H_m{}^1$ PLA melting enthalpy; $\Delta H_m{}^2$ BioPE melting enthalpy; X_c degree of crystallinity; T_{cc} cold crystallization temperature; ΔH_{cc} cold crystallization enthalpy; T_c melting crystallization temperature.

For PLA during heating, from the glass transition temperature (T_g) around 57–62 °C, an exothermic peak immediately before melting originating from the cold crystallization of the disordered α phase, then an endothermic peak ranging from 115.4 to 150.4 °C is observed in Figure 1A [34]. DSC scans of BioPE samples show the endothermic peak due to the melting with peak temperature at 136.8 °C, and the exothermic peak during cooling due to the melting crystallization with the temperature at 113.6 °C [35]. For PLA/BioPE samples, DSC scans displayed the exothermic and endothermic peaks characteristic of the individual polymers, whether compatibilized or not. The degree of crystallinity was evaluated through Equation (1); the results are presented in Table 3 together with associated parameters of the phase transitions.

PLA presented a low degree of crystallinity, i.e., X_c 3.9% whereas for BioPE high values were computed, i.e., X_c 60.1%. For bioblends, X_c increased, suggesting that the presence of BioPE increases the mobility of PLA chains, as well as the addition of compatibilizers, in response to the interactions developed between the macromolecular chains of PLA, BioPE and the compatibilizers, which tend to increase miscibility of samples [27].

Figure 2 illustrates the molten fraction and melting rate of the PLA-rich phase as temperature function. Parameters $T_{0.01}$, $T_{0.99}$ and C_{max} are shown in Table 4.

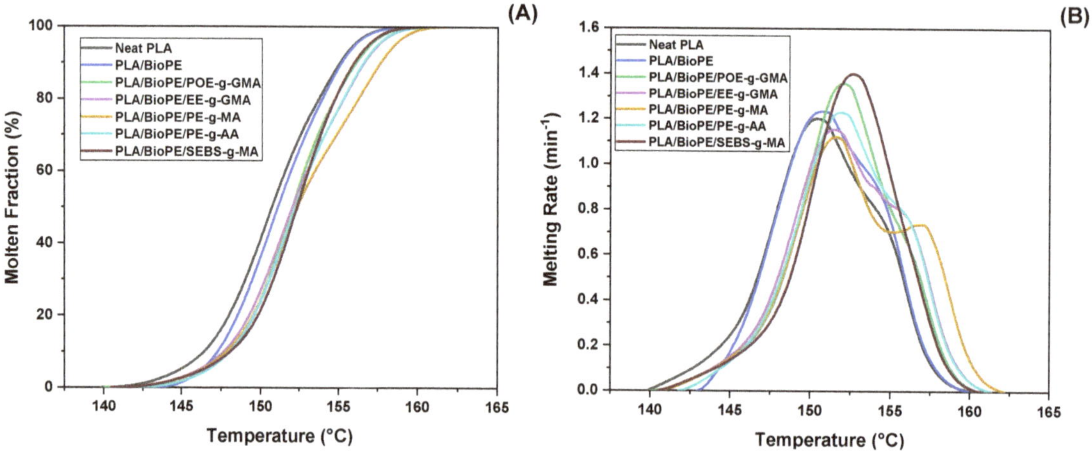

Figure 2. Molten fraction (**A**) and Melting rate (**B**) of investigated samples.

Table 4. Computed parameters from Figure 2 for the investigated samples.

Samples	$T_{0.01}$ (°C)	$T_{0.99}$ (°C)	C_{max} (min^{-1})
PLA	139.8	160.2	1.19
PLA/BioPE	142.9	159.9	1.22
PLA/BioPE/POE-g-GMA	140.4	160.4	1.35
PLA/BioPE/EE-g-GMA	140.6	161.1	1.15
PLA/BioPE/PE-g-MA	140.5	162.1	1.11
PLA/BioPE/PE-g-AA	141.8	161.3	1.21
PLA/BioPE/SEBS-g-MA	140.5	160.5	1.40

$T_{0.01}$ initial melting temperature; $T_{0.99}$ final melting temperature; C_{max} maximum melting rate.

Molten fraction displayed a sigmoidal character suggesting that the phase transition took place without discontinuities. In general, bioblends displayed subtle higher $T_{0.01}$ related to PLA. For $T_{0.99}$, there is a maintenance in relation to the neat PLA, with small increases for the compatibilized bioblends in relation to the binary bioblend. Regarding the melting rate Figure 2B, the bioblends compatibilized with POE-g-GMA and SEBS-g-MA showed the highest maximum melting rate (C_{max}) results both in relation to neat PLA and also to PLA/BioPE.

3.2. *Thermogravimetry (TG)*

Figure 3 presents TG plots of investigated samples and computed parameters from these plots are displayed in Table 5.

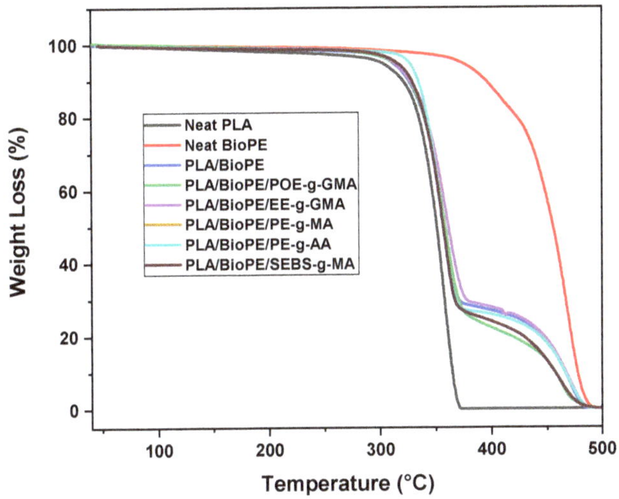

Figure 3. TG plots of investigated samples.

Table 5. $T_{D10\%}$, $T_{D50\%}$ and $T_{D99.9\%}$ parameters under inert atmosphere (N_2) and heating rate 10 °C/min.

Samples	$T_{D10\%}$ (°C)	$T_{D50\%}$ (°C)	$T_{D99.9\%}$ (°C)
PLA	320	350	373
BioPE	395	455	496
PLA/BioPE	326	358	485
PLA/BioPE/POE-g-GMA	327	358	490
PLA/BioPE/EE-g-GMA	328	361	487
PLA/BioPE/PE-g-MA	335	356	488
PLA/BioPE/PE-g-AA	335	356	487
PLA/BioPE/SEBS-g-MA	329	356	494

$T_{D10\%}$ = temperature for 10% of decomposed material. $T_{D50\%}$ = temperature for 50% of decomposed material. $T_{D99.9\%}$ = temperature for 99.9% of decomposed material.

From Figure 3 it is observed that neat polymers presented a single decomposition step, around 300–373 °C for PLA [36–38], and 350–496 °C for BioPE [35]. For PLA/BioPE bioblend, two decomposition steps were verified, where each decomposition step is characteristic of the individual polymer, i.e., PLA and BioPE. The compatibilized bioblends presented similar behavior to PLA/BioPE, however, as shown in Table 5, there was a subtle increase in $T_{D10\%}$ and $T_{D99.9\%}$, and small decrease in $T_{D50\%}$ for the compatibilized samples with PE-g-MA, PE-g-AA and SEBS-g-MA. It is worth noting that all bioblends had $T_{D10\%}$, $T_{D50\%}$ and $T_{D99.9\%}$ greater than neat PLA, with PLA/BioPE/PE-g-MA and PLA/BioPE/PE-g-AA bioblends having the highest values of $T_{D10\%}$ (335 °C), the bioblend PLA/BioPE/EE-g-GMA with the highest $T_{D50\%}$ (361 °C) and PLA/BioPE/SEBS-g-MA with the highest $T_{D99.9\%}$ (494 °C).

Therefore, gathered results indicate that the compatibilizing agents' addition to PLA/BioPE bioblend tends to improve the thermal stability not only in relation to PLA, but also to the binary bioblend, with improvement and/or maintenance of $T_{D10\%}$, $T_{D50\%}$ and $T_{D99.9\%}$ as reported.

3.3. Tensile Test

Elastic modulus data acquired under tension are illustrated in Figure 4. PLA and BioPE had an elastic modulus of 1221 and 376 MPa, respectively, with PLA having the highest stiffness [37,38], and BioPE increased flexibility, results corroborating these are shown later on for impact strength. PLA/BioPE due to the addition of 30% of a ductile material to PLA matrix, displayed decaying of 26.8% on the material's stiffness compared to PLA, i.e., it presented 893 MPa. In general, upon compatibilizing agent addition, the results were quite similar to PLA/BioPE, similar to those observed by Ferri et al. [27].

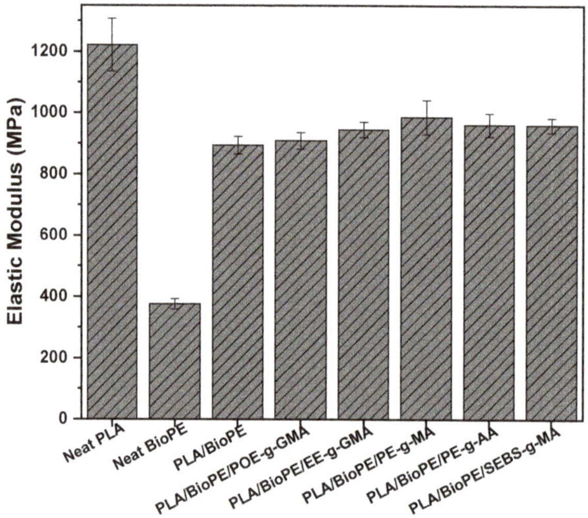

Figure 4. Elastic Modulus of investigated samples.

Figure 5 shows collected results for tensile strength. It is verified that PLA and BioPE have the highest and lowest tensile strength, respectively, i.e., 60 and 21 MPa, which are expected due to the fact that PLA has brittle material character requiring high stress to fracturing, while BioPE has ductile character. For PLA/BioPE bioblend (32 MPa), there was a considerable decrease compared to PLA, due to the addition of less rigid material, as well as due to poor adhesion between the phases of the system [39]. It may be observed that upon addition of the compatibilizers POE-g-GMA, EE-g-GMA, PE-g-MA, PE-g-AA and SEBS-g-MA, there was an increase in tensile strength, when compared to PLA/BioPE bioblends, for 38, 35, 39, 35 and 38 MPa, respectively, being directly linked to better adhesion between the phases present in the system, corroborating the impact strength results and the SEM images, which will be noted in later sections.

Figure 6 shows acquired results for the elongation at break of investigated samples. Differences are observed in relation to PLA/BioPE without and with compatibilizing agents. PLA/BioPE presented an elongation at break of 6.9%, bioblends compatibilized with POE-g-GMA, EE-g-GMA, PE-g-MA, PE-g-AA and SEBS-g-MA presented values of 9%, 8.7%, 8.3%, 8% and 10%, respectively. The increase in this property is related to the improvement in the stress transfer between the matrix and the dispersed phase of studied systems, due to the improvement in the interactions among the chemical groups present in the compatibilizers, showing its effectiveness [40,41]. These results corroborate those presented later on for impact strength, where the addition of POE-g-GMA, EE-g-GMA and SEBS-g-MA presented the best impact strengths, as well as higher elongations at break.

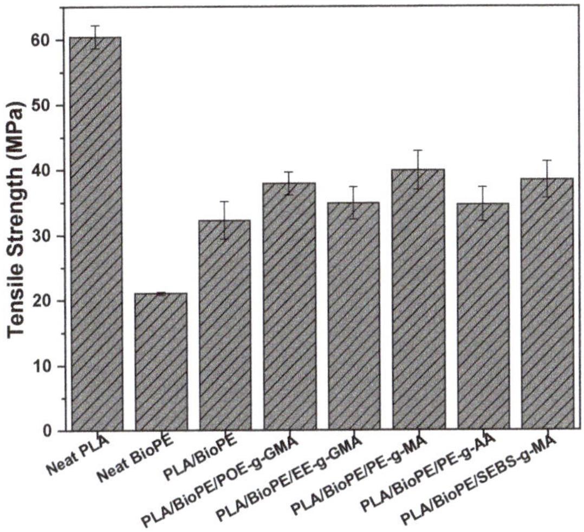

Figure 5. Tensile strength of investigated samples.

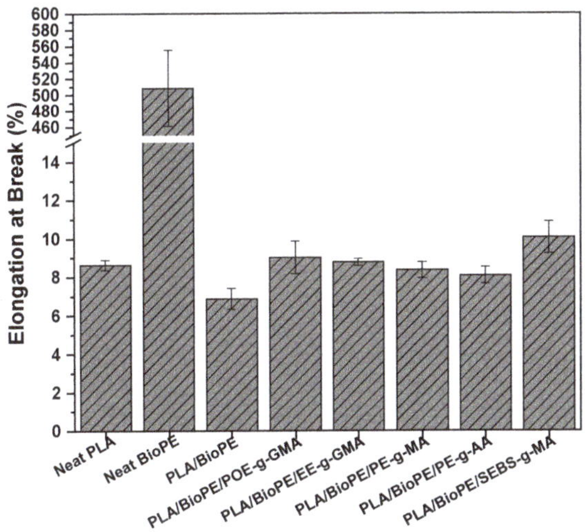

Figure 6. Elongation at break of investigated samples.

3.4. Impact Strength

Figure 7 illustrates the impact strength results obtained for the investigated samples. PLA displayed a typical fragile character with low energy dissipation with an impact strength of 27 J/m, which is in agreement with those already reported [12,14–16,27,38,42,43], and corroborated through SEM images later on presented (Figure 11). BioPE presented a typical ductile character with an impact strength of approximately 98 J/m [35]. Upon addition of 30% BioPE to PLA, there was an increase of 20% for PLA/BioPE related to PLA, even with the poor adhesion between the phases (see Figure 11).

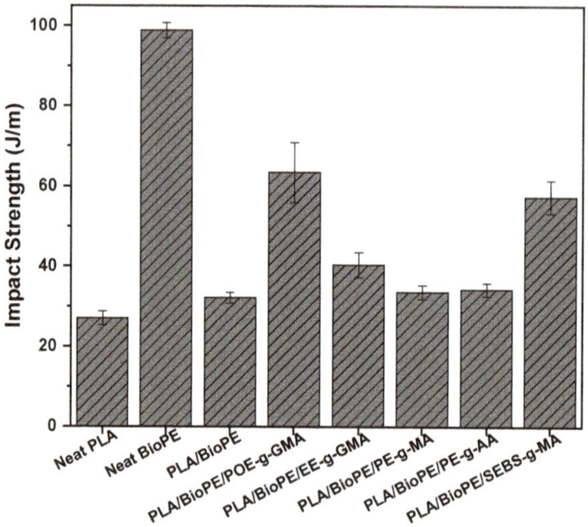

Figure 7. Impact strength of investigated samples.

For the compatibilized bioblends, impact strength increased with the addition of POE-g-GMA, EE-g-GMA and SEBS-g-MA, and remained unchanged with the addition of PE-g-MA and PE-g-AA. For compatibilizers grafted with GMA there was an increase of 100% for POE-g-GMA (~64 J/m) and approximately 30% for EE-g-GMA (~41 J/m), when compared to PLA/BioPE. This considerable improvement in toughness with the compatibilizers addition containing GMA is suggested to be related to chemical reactions between GMA epoxy groups and the terminal hydroxyl or carboxyl groups of PLA [24]. Upon SEBS-g-MA addition there was an increase of 78% compared to the bioblend without compatibilizer, with an impact strength of approximately 58 J/m, due to the fact that SEBS has elastomeric characteristic, and it is able to develop interactions between the phases with the elastomer grafted with MA, increasing the compound ductility.

Increases in impact strength were also observed in the reports of Ferri et al. [27] and Quiles et al. [44] in blends with PLA and PE, when added compatibilizers. Demonstrating the importance of adding a third phase to the immiscible PLA/BioPE system to improve properties such as toughness for example. Comparing the compatibilizers in relation to the degree of grafting of GMA, MA and AA, it may be verified that even using low levels of grafting of GMA (0.8%) for the compatibilizers POE-g-GMA and EE-g-GMA, compared to the highest MA contents (1.5–1.7 and 1.7%) for PE-g-MA and SEBS-g-MA, respectively, and AA (5.5–6.5%) for PE-g-AA, the predominant factor was the elastomeric characteristic of the used copolymers, with greater gains upon addition of POE-g-GMA and SEBS-g-MA.

The impact strength results of PLA/BioPE/POE-g-GMA and PLA/BioPE/SEBS-g-MA bioblends are relevant from a technological point of view, as they surpass the impact strength values of widely used commodity polymers in the development of products such as polypropylene (PP) [45–47], polystyrene (PS) [48,49] and polyamide 6 (PA6) [50–52]. From a sustainable point of view, it is extremely important, as polymers from renewable sources such as PLA and BioPE are being used in the bioblend, being an alternative for the use of the aforementioned polymers.

3.5. Heat Deflection Temperature (HDT)

The heat deflection temperature (HDT) becomes an important parameter for higher temperatures applications being an indication of the dimensional stability of the material under the effect of a specific load and temperature increasing [38,53].

Figure 8 presents HDT for the investigated samples. For PLA, HDT was 56 °C, a consequence of its glass transition temperature (T_g), as observed in the DSC scans [54,55].

BioPE had lower HDT (52 °C), being influenced, in this case, directly by the stiffness of the system, as observed in Figure 4, presenting the lowest value of elastic modulus among the studied samples. Upon addition of 30% BioPE to PLA matrix, HDT decayed only 1 °C compared to PLA. Upon addition of 10% of compatibilizer to PLA/BioPE, regardless of the compatibilizer, HDT decayed only 1 °C, related to PLA/BioPE. Therefore, comparing bioblends with PLA, there were no significant decreases in HDT, which is important from a technological and scientific point of view, an important parameter in the polymer industry, as well as linked to the significant increase in the material toughness as observed by the impact strength, especially for POE-g-GMA and SEBS-g-MA compatibilizers.

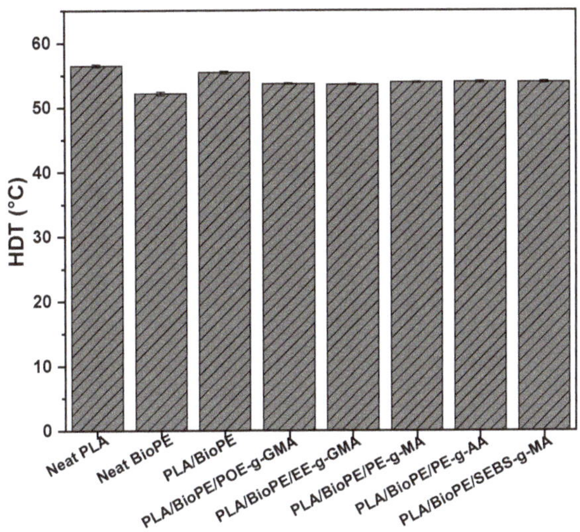

Figure 8. HDT of investigated samples.

3.6. Vicat Softening Temperature (VST) and Shore D Hardness

Similar to HDT, the Vicat softening temperature is considered an important property in polymeric systems, during production and projection of industrial applications aimed at thermomechanical resistance, being the temperature at which a needle penetrates the sample 1 mm, under a specific load, being mostly of materials, directly proportional to their surface hardness [53,56]. Figure 9 shows the relationship between VST and Shore D hardness of the investigated samples.

PLA presented VST around 59 °C, close to HDT. Upon temperature increase, the material goes from the solid to the rubbery state, hence reaching T_g, the needle penetrates the specimen. Shore D hardness testing is performed at room temperature; solid PLA has high surface hardness (69). Regarding BioPE, it displayed low hardness as it is a material with low stiffness, as observed in the tensile results, and high VST due to high crystallinity as verified through DSC [57].

For PLA/BioPE bioblend, Shore D hardness decreased whereas VST was unchanged, corroborating the decrease in stiffness provided by the BioPE phase, as also observed by Ferri et al. [27]. For the compatibilized bioblends, Shore D hardness results were similar to each other, being within the experimental error, while VST results decreased in relation to PLA/BioPE, with a reduction from 59 °C to 55–57 °C, probably due to T_g reduction, as seen in Table 3.

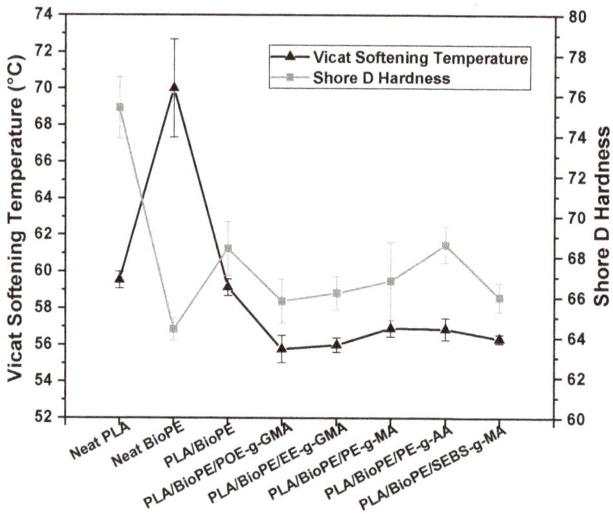

Figure 9. VST and Shore D Hardness of investigated samples.

3.7. Contact Angle

The contact angle results for the investigated materials are shown in Figure 10. The contact angle, in addition to enabling the assessment of the hydrophilicity or hydrophobicity of a material [38,58,59], also allows the assessment of surface energy [60], which is an important parameter during the production of compatible bioblends.

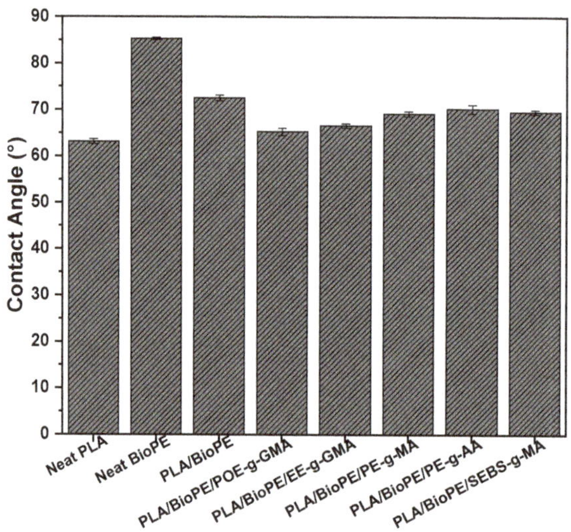

Figure 10. Contact angle of investigated samples.

Figure 10 shows the acquired data for the contact angle, where PLA displayed a value of 63°, and BioPE 85°, confirming greater hydrophobicity for BioPE, and PLA has chemical groups (hydroxyl and carboxyl), interacting with water, hence leading to smaller contact angle. PLA/BioPE bioblends presented an intermediate contact angle related to neat polymers, being approximately 72°. Upon compatibilizers addition, the contact angle decreased, in relation to PLA/BioPE bioblend, due to the increase in the interfacial and surface interactions, corroborating the results of impact strength and SEM. It is worth mentioning that chemical groups such as GMA, MA and AA present in the used compatibilizing agents, in addition to interacting with PLA and BioPE, tend to interact with water, increasing surface wettability and thus decreasing the contact angle of the systems in relation to PLA/BioPE bioblend.

3.8. Scanning Electron Microscopy (SEM)

SEM images of the specimens fractured surface from impact strength testing, of PLA, BioPE, PLA/BioPE and compatibilized bioblends are shown in Figure 11, with 500 and 1000× magnification for PLA, and 2000 and 5000× for the other samples.

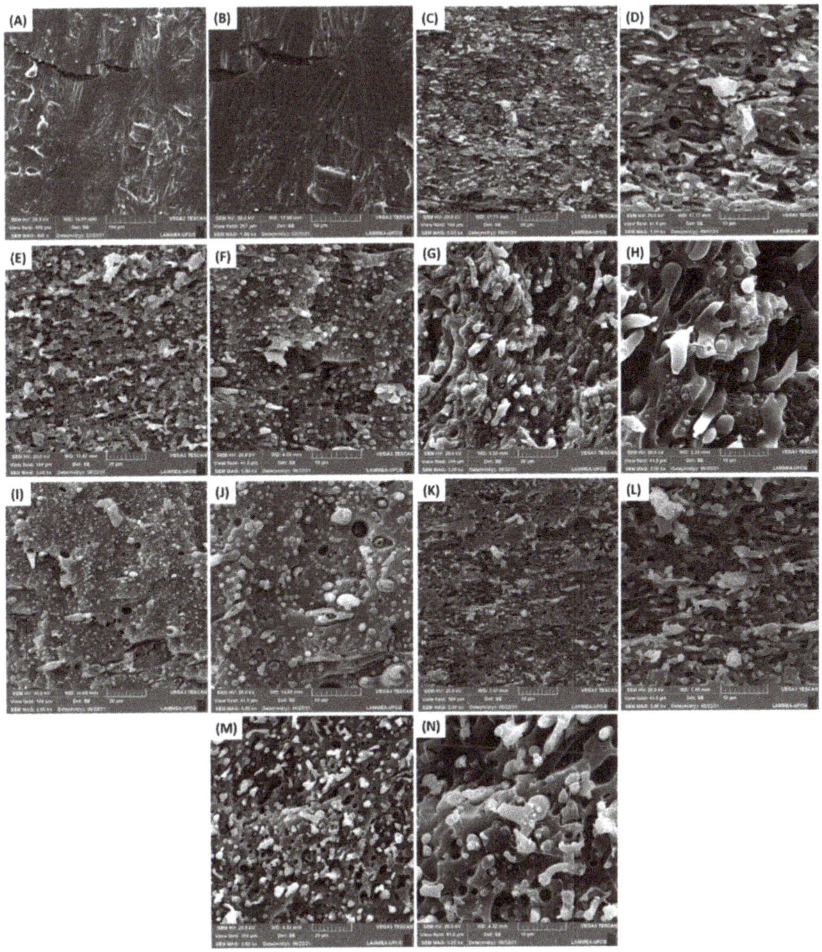

Figure 11. SEM images of PLA (**A**,**B**), PLA/BioPE (**C**,**D**), bioblends compatibilized with POE-g-GMA (**E**,**F**), EE-g-GMA (**G**,**H**), PE-g-MA (**I**,**J**), PE-g-AA (**K**,**L**) and SEBS-g-MA (**M**,**N**). Images are 500 and 1000× magnification of neat PLA, and 2000 and 5000× for the other compositions.

Regarding PLA (Figure 11a,b), smooth surface without roughness was observed, due to the absence of plastic deformation, characteristic of fragile fracture, as observed in the impact strength and tensile experiments [37,38,61–63].

SEM images of PLA/BioPE without compatibilizer showed immiscibility and poor adhesion between PLA and BioPE polymers, with BioPE particles pulled out from PLA matrix, with presence of voids in the fracture surface, promoting low tensile and impact strength properties [24,27].

For the compatibilized bioblends, the roughness was verified on the specimens' surfaces, characteristic of plastic deformation, mainly for PLA/BioPE/POE-g-GMA, PLA/BioPE/EE-g-GMA and PLA/BioPE/SEBS-g-MA, due to the elastomeric character present in the copolymer, resulting in more elongated particles after the impact test. Related to the addition of PE-g-MA and PE-g-AA compatibilizers, there was lower surface roughness, with homogeneity when compared to the PLA/BioPE bioblend, due to the presence of MA and AA groups, resulting in improvement in interactions as well as good tensile strength results, mainly for PE-g-AA, but with the maintenance of impact strength compared to PLA/BioPE bioblend. Thus, it was observed that the elastomeric character of the copolymers was predominant in the best impact strength results compared to PLA/BioPE bioblend, that is, the compatibilizers POE-g-GMA, EE-g-GMA, and SEBS-g-GMA obtained the best impact strength results, not only due to the improved interactions between PLA and GMA and MA groups, but also due to the elastomeric character of these copolymers.

4. Conclusions

The addition effect of POE-g-GMA, EE-g-GMA, PE-g-MA, PE-g-AA and SEBS-g-MA on the thermal, mechanical, thermomechanical properties, wettability (contact angle) and morphology of PLA/BioPE bioblends was investigated. Regarding thermal analysis, it was observed by DSC that the BioPE addition as well as compatibilizing agents increase the degree of crystallinity of neat PLA, increase the macromolecular chains, and provide significant improvement in thermal stability, mainly in parameter $T_{D99.9\%}$, as observed by TG. The mechanical and thermomechanical properties demonstrated that the compatibilizing agents improve the interactions among bioblends phases, with property increases such as elongation at break, tensile strength, impact strength, and unchanged elastic modulus, related to PLA/BioPE. Through contact angle measurements, an increase in wettability was observed compared to PLA/BioPE. SEM images showed that the addition of POE-g-GMA, EE-g-GMA and SEBS-g-MA considerably increased the roughness, promoting plastic deformation and better impact strength performance related to PLA/BioPE and PLA. The addition of POE-g-GMA compatibilizer to PLA/BioPE displayed the best overall balance in relation to the investigated properties. Acquired results indicate that PLA/BioPE needs to be compatibilized in order to promote interactions among phases, and consequently improve the technological performance. The manufacture of bioblends is an alternative for the commercialization of eco-products that contribute to a friendlier environmental cycle.

Author Contributions: All authors participated in the drafting the article or revising it critically content, approving the final version submitted. Conceptualization, E.d.S.B.F., C.B.B.L., E.M.A. and R.M.R.W.; methodology, E.d.S.B.F., D.D.S., and E.A.d.S.F.; formal analysis, E.d.S.B.F., C.B.B.L. and E.A.d.S.F.; investigation, E.d.S.B.F., C.B.B.L. and D.D.S.; resources, E.M.A.; writing—original draft preparation, E.d.S.B.F. and C.B.B.L.; writing—review and editing, E.d.S.B.F. and E.M.A. and R.M.R.W.; visualization, E.d.S.B.F. and C.B.B.L.; supervision, E.M.A.; project administration, E.M.A. and R.M.R.W.; funding acquisition, E.M.A. All authors have read and agreed to the published version of the manuscript.

Funding: The authors are grateful to CNPq (National Council for Scientific and Technological Development, Brasilia/DF, Brazil) and CAPES (Coordination for the Improvement of Higher Education Personnel, Brasilia/DF, Brazil) for the financial support. Edcleide Araújo and Renate Wellen are CNPq fellows.

Acknowledgments: The author would to thank Addivant for PE-g-MA and PE-g-AA donation; Xiamen Coace Plastic Technology for POE-g-GMA and EE-g-GMA donation and Kraton Polymers Group of Companies for SEBS-g-MA donation and FAPESQ for support. The authors thank UFCG for paying the publication fee.

Conflicts of Interest: There is no conflict of interest and all authors have agreed with this submission and they are aware of the content.

References

1. Mohamed, A.; Gordon, S.H.; Biresaw, G. Polycaprolactone/polystyrene bioblends characterized by thermogravimetry, modulated differential scanning calorimetry and infrared photoacoustic spectroscopy. *Polym. Degrad. Stab.* **2007**, *92*, 1177–1185. [CrossRef]
2. Al-Itry, R.; Lamnawar, K.; Maazouz, A. Biopolymer Blends Based on Poly (lactic acid): Shear and Elongation Rheology/Structure/Blowing Process Relationships. *Polymers* **2015**, *7*, 939–962. [CrossRef]
3. Olejnik, O.; Masek, A.; Zawadziłło, J. Processability and Mechanical Properties of Thermoplastic Polylactide/Polyhydroxybutyrate (PLA/PHB) Bioblends. *Materials* **2021**, *14*, 898. [CrossRef]
4. Kim, Y.F.; Choi, C.N.; Lee, K.Y.; Lee, M.S. Compatibilization of immiscible poly(l-lactide) and low density polyethylene blends. *Fibers Polym.* **2004**, *5*, 270–274. [CrossRef]
5. La Mantia, F.P.; Morreale, M.; Botta, L.; Mistretta, M.; Ceraulo, M.; Scaffaro, R. Degradation of polymer blends: A brief review. *Polym. Degrad. Stab.* **2017**, *145*, 79–92. [CrossRef]
6. Effah, B.; Van Reenen, A.; Meincken, M. Mechanical properties of wood-plastic composites made from various wood species with different compatibilizers. *Eur. J. Wood Wood Prod.* **2018**, *76*, 57–68. [CrossRef]
7. Luna, C.B.B.; Siqueira, D.D.; Araújo, E.M.; Wellen, R.M.R. Tailoring PS/PPrecycled blends compatibilized with SEBS. Evaluation of rheological, mechanical, thermomechanical and morphological characters. *Mater. Res. Express* **2019**, *6*, 075316. [CrossRef]
8. Czarnecka-Komorowska, D.; Nowak-Grzebyta, J.; Gawdzińska, K.; Mysiukiewicz, O.; Tomasik, M. Polyethylene/Polyamide Blends Made of Waste with Compatibilizer: Processing, Morphology, Rheological and Thermo-Mechanical Behavior. *Polymers* **2021**, *13*, 2385. [CrossRef]
9. Ferreira, E.D.S.B.; Luna, C.B.B.; Araújo, E.M.; Siqueira, D.D.; Wellen, R.M.R. Polypropylene/wood powder/ethylene propylene diene monomer rubber-maleic anhydride composites: Effect of PP melt flow index on the thermal, mechanical, thermomechanical, water absorption, and morphological parameters. *Polym. Compos.* **2021**, *42*, 484–497. [CrossRef]
10. Silva, P.T.V.; Luna, C.B.B.; Ferreira, E.S.B.; Santos Filho, E.A.; Araújo, E.M. Evolution of morphology and mechanical properties of PA6/AES blends compatibilized with EPDM-MA. *Res. Soc. Dev.* **2021**, *10*, e210101018791. [CrossRef]
11. Nogueira, J.A.S.; Luna, C.B.B.; Siqueira, D.D.; Santos Filho, E.A.; Araújo, E.M. Performance investigation of PA6/HIPS blends compatibilized with SEBS-MA. Effect of compatibilizer content on mechanical, thermomechanical, torque rheometry and morphology properties. *Res. Soc. Dev.* **2021**, *10*, e58510817649.
12. Saini, P.; Arora, M.; Kumar, M.R. Poly(lactic acid) blends in biomedical applications. *Adv. Drug Deliv. Rev.* **2016**, *107*, 47–59. [CrossRef] [PubMed]
13. Ye, G.; Wang, W.; Fan, D.; He, P. Effects of femtosecond laser micromachining on the surface and substrate properties of poly-lactic acid (PLA). *Appl. Surf. Sci.* **2021**, *538*, 148117. [CrossRef]
14. Garlotta, D. A Literature Review of Poly(Lactic Acid). *J. Polym. Environ.* **2001**, *9*, 63–84. [CrossRef]
15. Gigante, V.; Canesi, I.; Cinelli, P.; Coltelli, M.B.; Lazzeri, A. Rubber Toughening of Polylactic Acid (PLA) with Poly(butylene adipate-co-terephthalate) (PBAT): Mechanical Properties, Fracture Mechanics and Analysis of Ductile-to-Brittle Behavior while Varying Temperature and Test Speed. *Eur. Polym. J.* **2019**, *115*, 125–137. [CrossRef]
16. Wu, F.; Misra, M.; Mohanty, A.K. Super Toughened Poly(lactic acid)-Based Ternary Blends via Enhancing Interfacial Compatibility. *ACS Omega* **2019**, *4*, 1955–1968. [CrossRef]
17. McKeown, P.; Jones, M.D. The Chemical Recycling of PLA: A Review. *Sustain. Chem.* **2020**, *1*, 1. [CrossRef]
18. Boronat, T.; Fombuena, V.; Garcia-Sanoguera, D.; Sanchez-Nacher, L.; Balart, R. Development of a biocomposite based on green polyethylene biopolymer and eggshell. *Mater. Des.* **2015**, *68*, 177–185. [CrossRef]
19. Spalding, M.A.; Chatterjee, A. *Handbook of Industrial Polyethylene and Technology: Definitive Guide to Manufacturing, Properties, Processing, Applications and Markets*; John Wiley & Sons: Hoboken, NJ, USA, 2017.
20. Mazur, K.; Jakubowska, P.; Romańska, P.; Kuciel, S. Green high density polyethylene (HDPE) reinforced with basalt fiber and agricultural fillers for technical applications. *Compos. Part B Eng.* **2020**, *202*, 108399. [CrossRef]
21. Anderson, K.S.; Hillmyer, M.A. The influence of block copolymer microstructure on the toughness of compatibilized polylactide/polyethylene blends. *Polymer* **2004**, *45*, 8809–8823. [CrossRef]
22. Thurber, C.M.; Xu, Y.; Myers, J.C.; Lodge, T.P.; Macosko, C.W. Accelerating Reactive Compatibilization of PE/PLA Blends by an Interfacially Localized Catalyst. *ACS Macro Lett.* **2014**, *4*, 30–33. [CrossRef]
23. Xu, Y.; Loi, J.; Delgado, P.; Topolkaraev, V.; McEneany, R.J.; Macosko, C.W.; Hillmyer, M.A. Reactive Compatibilization of Polylactide/Polypropylene Blends. *Ind. Eng. Chem. Res.* **2015**, *54*, 6108–6114. [CrossRef]
24. Brito, G.F.; Agrawal, P.; Mélo, T.J.A.; Pinto, J.C.; Balaban, R. Mechanical and Morphological Properties of PLA/BioPE Blend Compatibilized with E-GMA and EMA-GMA Copolymers. *Macromol. Symp.* **2016**, *367*, 176–182. [CrossRef]
25. Quitadamo, A.; Massardier, V.; Santulli, C.; Valente, M. Optimization of Thermoplastic Blend Matrix HDPE/PLA with Different Types and Levels of Coupling Agents. *Materials* **2018**, *11*, 2527. [CrossRef] [PubMed]
26. Zolali, A.M.; Favis, B.D. Toughening of cocontinuous polylactide/polyethylene blends via na interfacially percolated intermediate phase. *Macromolecules* **2018**, *51*, 3572–3581. [CrossRef]
27. Ferri, J.M.; Garcia-Garcia, D.; Rayón, E.; Samper, M.D.; Balart, R. Compatibilization and Characterization of Polylactide and Biopolyethylene Binary Blends by Non-Reactive and Reactive Compatibilization Approaches. *Polymers* **2020**, *12*, 1344. [CrossRef] [PubMed]

28. Srithep, Y.; Nealey, P.; Turng, L.-S. Effects of annealing time and temperature on the crystallinity and heat resistance behavior of injection-molded poly(lactic acid). *Polym. Eng. Sci.* **2013**, *53*, 580–588. [CrossRef]
29. Wang, S.; Liu, B.; Qin, Y.; Guo, H. Effects of Processing Conditions and Plasticizing-Reinforcing Modification on the Crystallization and Physical Properties of PLA Films. *Membranes* **2021**, *11*, 640. [CrossRef]
30. Guo, H.; Rinaldi, R.G.; Tayakout, S.; Broudin, M.; Lame, O. The correlation between the mixed-mode oligo-cyclic loading induced mechanical and microstructure changes in HDPE. *Polymer* **2021**, *224*, 123706. [CrossRef]
31. Wellen, R.M.; Rabello, M.S.; Fechine, G.J.; Canedo, E.L. The melting behaviour of poly (3-hydroxybutyrate) by DSC. Reproducibility study. *Polym. Test.* **2013**, *32*, 215–220. [CrossRef]
32. Luna, C.B.B.; Ferreira, E.D.S.B.; Siqueira, D.D.; Filho, E.A.D.S.; Araújo, E.M. Additivation of the ethylene–vinyl acetate copolymer (EVA) with maleic anhydride (MA) and dicumyl peroxide (DCP): The impact of styrene monomer on cross-linking and functionalization. *Polym. Bull.* **2021**, *78*, 1–24. [CrossRef]
33. Barros, A.B.D.S.; Farias, R.D.F.; Siqueira, D.D.; Luna, C.B.B.; Araújo, E.M.; Rabello, M.S.; Wellen, R.M.R. The Effect of ZnO on the Failure of PET by Environmental Stress Cracking. *Materials* **2020**, *13*, 2844. [CrossRef]
34. Mohapatra, A.K.; Mohanty, S.; Nayak, S. Poly(lactic acid) and layered silicate nanocomposites prepared by melt mixing: Thermomechanical and morphological properties. *Polym. Compos.* **2012**, *33*, 2095–2104. [CrossRef]
35. da Silva, F.S.; Luna, C.B.B.; Siqueira, D.D.; Ferreira, E.D.S.B.; Araújo, E.M. From Waste to Reuse: Manufacture of Ecological Composites Based on Biopolyethylene/wood Powder with PE-g-MA and Macaíba Oil. *J. Polym. Environ.* **2021**, *29*, 1–17. [CrossRef]
36. Abdolrasouli, M.H.; Sadeghi, G.M.M.; Nazockdast, H.; Babaei, A. Polylactide/Polyethylene/Organoclay Blend Nanocomposites: Structure, Mechanical and Thermal Properties. *Polym. Technol. Eng.* **2014**, *53*, 1417–1424. [CrossRef]
37. da Silva, W.A.; Luna, C.B.B.; Melo, J.B.D.C.A.D.; Araújo, E.M.; Filho, E.A.D.S.; Duarte, R.N.C. Feasibility of Manufacturing Disposable Cups using PLA/PCL Composites Reinforced with Wood Powder. *J. Polym. Environ.* **2021**, *29*, 2932–2951. [CrossRef]
38. Ferreira, E.S.B.; Luna, C.B.B.; Siqueira, D.D.; Araújo, E.M.; De França, D.C.; Wellen, R.M.R. Annealing Effect on PLA/EVA Blends Performance. *J. Polym. Environ.* **2021**, *29*, 1–14.
39. Vrsaljko, D.; Macut, D.; Kovačević, V. Potential role of nanofillers as compatibilizers in immiscible PLA/LDPE Blends. *J. Appl. Polym. Sci.* **2015**, *132*, 132. [CrossRef]
40. Pai, F.-C.; Chu, H.-H.; Lai, S.-M. Reactive compatibilization of poly(lactic acid)/polyethylene octene copolymer blends with ethylene-glycidyl methacrylate copolymer. *J. Polym. Eng.* **2011**, *31*, 463–471. [CrossRef]
41. Gallego, R.; López-Quintana, S.; Basurto, F.; Núñez, K.; Villarreal, N.; Merino, J.C. Synthesis of new compatibilizers to poly (lactic acid) blends. *Polym. Eng. Sci.* **2014**, *54*, 522–530. [CrossRef]
42. Brito, G.F.; Agrawal, P.; Araújo, E.M.; De Melo, T.J.A. Polylactide/Biopolyethylene bioblends. *Polímeros* **2012**, *22*, 427–429. [CrossRef]
43. Huang, J.; Fan, J.; Yuan, D.; Zhang, S.; Chen, Y. Facile preparation of super-toughened polylactide-based thermoplastic vulcanizates without sacrificing the stiffness based on the selective distribution of silica. *Ind. Eng. Chem. Res.* **2020**, *59*, 9950–9958. [CrossRef]
44. Quiles-Carrillo, L.; Montanes, N.; Jorda-Vilaplana, A.; Balart, R.; Torres-Giner, S. A comparative study on the effect of different reactive compatibilizers on injection-molded pieces of bio-based high-density polyethylene/polylactide blends. *J. Appl. Polym. Sci.* **2019**, *136*, 47396. [CrossRef]
45. Nascimento, W.A.; Silva, H.P.D.S.; Agrawal, P.; Lima, J.C.; de Melo, T.J.; Lira, H.D.L. Correlação entre propriedades mecânicas, reológicas e morfológicas de blendas PP/PA6 com adição de compatibilizantes. *Rev. Iberoam. Polim.* **2020**, *21*, 144–154.
46. Pruthtikul, R.; Liewchirakorn, P. Preparation of Polypropylene Graft Maleic Anhydride (PP-g-MA) via Twin Screw Extrusion. *Adv. Mater. Res.* **2010**, *93–94*, 451–454. [CrossRef]
47. Mahesh, A.; Rudresh, B.; Reddappa, H. Potential of natural fibers in the modification of mechanical behavior of polypropylene hybrid composites. *Mater. Today Proc.* **2021**, *46*, 1–6. [CrossRef]
48. Luna, C.; Siqueira, D.; Ferreira, E.; Silva, W.; Nogueira, J.; Araújo, E. From Disposal to Technological Potential: Reuse of Polypropylene Waste from Industrial Containers as a Polystyrene Impact Modifier. *Sustainability* **2020**, *12*, 5272. [CrossRef]
49. Morais, D.D.D.S.; França, D.C.; Araújo, E.M.; Carvalho, L.H.D.; Wellen, R.M.R.; Oliveira, A.D.D.; Melo, T.J.A.D. Tayloring PS/PCL blends: Characteristics of processing and properties. *REM-Int. Eng. J.* **2019**, *72*, 87–95. [CrossRef]
50. Castro, L.; Oliveira, A.; Kersch, M.; Altstädt, V.; Pessan, L. Effects of mixing protocol on morphology and properties of PA6/ABS blends compatibilized with MMA-MA. *J. Appl. Polym. Sci.* **2016**, *133*, 1–8. [CrossRef]
51. Oliveira, A.D.D.; Larocca, N.M.; Pessan, L.A. Effect from the blending sequence on the properties of PA6/ABS blends compatibilized with SMA copolymer. *Polímeros* **2011**, *21*, 27–33. [CrossRef]
52. De Oliveira, A.D.; De Castro, L.D.C.; Beatrice, C.A.G.; Lucas, A.D.A.; Pessan, L.A. Effect of Maleic Anhydride Content in Properties of PA6/AES Blends Compatibilized with MMA-MA. *Mater. Res.* **2017**, *20*, 1630–1637. [CrossRef]
53. Luna, C.B.B.; Siqueira, D.D.; Ferreira, E.D.S.B.; Araújo, E.M.; Wellen, R.M.R. Reactive compatilization of PCL/WP upon addition of PCL-MA. Smart option for recycling industry. *Mater. Res. Express* **2019**, *6*, 125317. [CrossRef]
54. Wang, L.; Wang, Y.-N.; Huang, Z.-G.; Weng, Y.-X. Heat resistance, crystallization behavior, and mechanical properties of polylactide/nucleating agent composites. *Mater. Des.* **2015**, *66*, 7–15. [CrossRef]

55. Yang, Y.; Zhang, L.; Xiong, Z.; Tang, Z.; Zhang, R.; Zhu, J. Research progress in the heat resistance, toughening and filling modification of PLA. *Sci. China Ser. B Chem.* **2016**, *59*, 1355–1368. [CrossRef]
56. Ferreira, E.D.S.B.; Luna, C.B.B.; Araújo, E.M.; Siqueira, D.D.; Wellen, R.M.R. Polypropylene/wood powder composites: Evaluation of PP viscosity in thermal, mechanical, thermomechanical, and morphological characters. *J. Thermoplast. Compos. Mater.* **2019**, *32*, 0892705719880958. [CrossRef]
57. Tao, Y.U.; Yan, L.I.; Jie, R.E.N. Preparation and properties of short natural fiber reinforced poly (lactic acid) composites. *Trans. Nonferrous Met. Soc. China* **2009**, *19*, 651–655.
58. Bezerra, E.B.; De França, D.C.; Morais, D.D.D.S.; Silva, I.D.D.S.; Siqueira, D.D.; Araújo, E.M.; Wellen, R.M.R. Compatibility and characterization of Bio-PE/PCL blends. *Polímeros* **2019**, *29*, 1–15. [CrossRef]
59. Siqueira, D.; Luna, C.; Ferreira, E.; Araújo, E.; Wellen, R. Tailored PCL/Macaíba fiber to reach sustainable biocomposites. *J. Mater. Res. Technol.* **2020**, *9*, 9691–9708. [CrossRef]
60. Siročić, A.P.; Hrnjak-Murgić, Z.; Jelenčić, J. The surface energy as an indicator of miscibility of SAN/EDPM polymer blends. *J. Adhes. Sci. Technol.* **2013**, *27*, 2615–2628. [CrossRef]
61. Tejada-Oliveros, R.; Balart, R.; Ivorra-Martinez, J.; Gomez-Caturla, J.; Montanes, N.; Quiles-Carrillo, L. Improvement of Impact Strength of Polylactide Blends with a Thermoplastic Elastomer Compatibilized with Biobased Maleinized Linseed Oil for Applications in Rigid Packaging. *Molecules* **2021**, *26*, 240. [CrossRef]
62. Moradi, S.; Yeganeh, J.K. Highly toughened poly(lactic acid) (PLA) prepared through melt blending with ethylene-co-vinyl acetate (EVA) copolymer and simultaneous addition of hydrophilic silica nanoparticles and block copolymer compatibilizer. *Polym. Test.* **2020**, *91*, 106735. [CrossRef]
63. Sangeetha, V.H.; Valapa, R.B.; Nayak, S.K.; Varghese, T.O. Investigation on the Influence of EVA Content on the Mechanical and Thermal Characteristics of Poly(lactic acid) Blends. *J. Polym. Environ.* **2016**, *26*, 1–14. [CrossRef]

sustainability

MDPI

Article

Durability Behavior of Mortars Containing Perlite Tailings: Alkali–Silicate Reaction Viewpoint

Roberto Evaristo de Oliveira Neto [1], Juliana de Melo Cartaxo [2], Alisson Mendes Rodrigues [2,*], Gelmires de Araújo Neves [2], Romualdo Rodrigues Menezes [2], Fabiana Pereira da Costa [1] and Sâmea Valensca Alves Barros [3]

1 Materials Science and Engineering, Federal University of Campina Grande, Av. Aprígio Veloso-882, Bodocongó, Campina Grande 58429-900, PB, Brazil; mscreon@gmail.com (R.E.d.O.N.); fabiana.costa@estudante.ufcg.edu.br (F.P.d.C.)

2 Laboratory of Materials Technology (LTM), Department of Materials Engineering, Federal University of Campina Grande (UFCG), Av. Aprígio Veloso-882, Bodocongó, Campina Grande 58429-900, PB, Brazil; julianamelo25@gmail.com (J.d.M.C.); gelmires.neves@ufcg.edu.br (G.d.A.N.); romualdo.menezes@ufcg.edu.br (R.R.M.)

3 Engineering Department, Universidade Federal Rural de Semi-Árido, Rua Gamaliel Martins Bezerra, Alto da Alegria, Angicos 59515-000, RN, Brazil; sameavalensca@ufersa.edu.br

* Correspondence: alisson.mendes@professor.ufcg.edu.br

Abstract: Tailing incorporation into mortars has been the subject of much research in recent years. Despite this, most of these studies did not investigate the harmful effects resulting from the exposure of such mortars to an environment containing aggressive agents. This work investigated the effects of perlite tailing addition into mortars containing cement CP V-ARI MAX and hydrated lime. The raw materials were subjected to chemical characterization (X-ray fluorescence (XRF)) and mineralogical (X-ray diffraction (XRD)), while the samples immersed in 1 N NaOH solution were characterized by XRD, thermogravimetric analysis (TGA), scanning electron microscopy (SEM), and compression strength (CS). The results showed the harmful effects of incorporating perlite tailings into the mortar investigated. Such a degradation was proven by linear expansion and compressive strength experiments accomplished in the samples after the test of resistance to an alkali–silicate reaction.

Keywords: perlite tailings; aggregate; alkali–silicate reaction

Citation: Neto, R.E.d.O.; Cartaxo, J.d.M.; Rodrigues, A.M.; Neves, G.d.A.; Menezes, R.R.; da Costa, F.P.; Barros, S.V.A. Durability Behavior of Mortars Containing Perlite Tailings: Alkali–Silicate Reaction Viewpoint. *Sustainability* **2021**, *13*, 9203. https://doi.org/10.3390/su13169203

Academic Editor: Antonio Caggiano

Received: 20 July 2021
Accepted: 13 August 2021
Published: 17 August 2021

1. Introduction

Perlite is a natural vitreous volcanic rock that, when heated to around 900–1200 °C, expands 5 to 20 times its original volume [1], making it suitable for various applications in the most diverse industries, such as construction, paint industries, pharmaceuticals, soil and water filtration, and the plastics industry [2–5]. However, during the processing of raw pearlite (i.e., cutting, crushing, grinding, and particle size classification), a large volume of tailings is generated, with particle sizes ranging from microns to about 0.5 mm [6]. In this processing process, two types of waste are generated: the coarse perlite (CP) tailings from the crushing stages and the fine tailings (PP) from the grinding and granulometric classification stages. The expanded perlite production process also generates a significant amount of tailings; these are finely granulometry (<100 μm) and extremely low bulk density (50–150 kg/m^3), which makes them difficult to handle [7]. These tailings, whether natural or expanded, are often improperly disposed of, causing negative impacts on the environment [8–11].

The need to give an adequate destination to perlite tailings has generated numerous researches which using these tailings for different purposes [12–17]. The technical feasibility of using the tailings of expanded perlite in Portland cement mortar was studied regarding compression strength and proven by the study of Kapelusznaet et al. [8]. Erdoğan and Sağlık [18] investigated the use of natural perlite tailings (non-expanded) in mortars as a partial replacement for cement (25% and 50%). The authors reported that incorporating 25 and 50% of the perlite tailings decreased the compressive strength by 4.45% and 32.29%, respectively.

Ramezanianpour et al. [14] incorporated calcined perlite powder (1 h at 850 °C), finely ground (fineness of 3100 cm^2/g) as supplementary cementitious material (SCMs) in concrete. The authors reported that the incorporation of fine calcined perlite powders consumes lime and reduces the conductivity of the cement pore solution due to pozzolanic reactions. Most studies have focused on using expanded perlite tailings mainly as pozzolanic material [19,20]. Therefore, it is noted that there is a lack of studies related to the use of natural (non-expanded) perlite tailings. Furthermore, there is a gap in the literature about the harmful potential of these tailings concerning alkali-aggregate reactions (AARs).

Alkali-aggregate reactions occur between alkalis (Na_2O and K_2O) present in Portland cement and reactive aggregates in the presence of high moisture (80–85%) and adequate temperature beyond the pH [21,22]. An expansive gel is among the main products of AARs. Such a gel absorbs water, swells, and leads to the appearance of cracks and consequent collapse of the cement matrix [22–24]. AARs can be classified into three types: alkali–silica, alkali–carbonate, and alkali–silicate reactions [25]. The alkali–silica reaction occurs between the alkalis in the cement and the reactive minerals in the aggregate. Meanwhile, the alkaline–carbonate reaction occurs between certain dolomitic limestones and the alkaline solutions present in the pores of the concrete. On the other hand, the alkali–silicate reaction has the same nature as the alkali–silica reaction but with a slower process involving some silicates present in feldspars, certain metamorphic and sedimentary igneous rocks, deformed (strained) quartz, and expansive minerals. Furthermore, when alkali–aggregate reactions occur in cementitious matrices, they cannot be interrupted; thus, preventing the occurrence of these reactions is the best way to produce mortars with durability to chemical attacks [25,26].

In this context, there is a clear need to investigate the aggregates used in mortar and concrete because, even if types of cement with low alkali concentration are used, reactions can still occur because the aggregates can be constituted by reactive minerals formed by alkali silicates, which may react with the hydroxyl ions released by portlandite during the cement hydration process, causing the alkali–silicate reaction (ASR). It is known that the perlite tailings contain sodium and potassium aluminum silicate phases (i.e., anorthoclase, (Na, K) $AlSi_3O_8$) [27]. The reactive minerals in the anorthoclase react with the alkalis present in Portland cement, triggering the alkali–silicate reaction.

Therefore, it is necessary to investigate the harmful potential of perlite tailings in relation to AARs to be safely used as an aggregate in mortar and concrete without compromising their duration of usefulness. Thus, the objective of this work is to determine the harmful potential of tailings from perlite processing in relation to alkali–silicate reactions. For this, the perlite tailings were chemically and mineralogically characterized (X-ray fluorescence and X-ray diffraction), while evidence after the attack in a solution of 1 N NaOH was characterized by XRD, TGA, SEM, and CS.

2. Materials and Methods

2.1. Materials

The materials used in this research were cement CP V-ARI MAX, hydrated lime (Carbomil S.A., Limoeiro do Norte, CE, Brazil), and tailings from the processing of perlite (Bentonita do Nordeste S.A., PB, Brazil). The perlite tailings were received in two distinct granulometry: powdered and coarse perlite (PP e CP, respectively). The cement CP V-ARI MAX was used in this research because it has low alkalinity (0.69%) and does not have pozzolanic material in its composition; this makes it possible to investigate the influence of mineral tailings additions in the expansion process. According to information from the manufacturer, the cement CP V-ARI MAX has ~70% clinker. Calcitic lime was used because it has a content of CaO greater than 95%. The chemical composition of cement CP V-ARI and lime are shown in Table 1.

Table 1. Chemical composition of the CPV-ARI MAX cement and lime used in this research.

Raw Material	Oxides (%)										
	CaO	SiO_2	MgO	Al_2O_3	Fe_2O_3	SO_3	K_2O	TiO_2	Na_2O_{eq} [a]	LOI [b]	C_3A [c]
Cement	61.35	19.1	1.33	5.33	4.1	4.87	1.04	0.5	0.69	2.2	7.18
Lime	62.54	0.56	1.20	0.27	0.15	—	0.25	—	—	35.0	—

[a] Na_2O_{eq} = %Na_2O + 0.658(%K_2O); [b] LOI: loss on ignition; [c] %C_3A = 2.650(%Al_2O_3) − 1.692(%Fe_2O_3).

The formulations cement-containing mortars, lime, and perlite tailings (PP and CP) investigated in this work were previously developed by our research group [27]. The mortar traces used were 1:3 (cement: 100% CP); 1:3 (cement: 85% CP + 15% PP); 1:1:6 (cement: lime: 100% CP), and 1:1:6 (cement: lime: 85% CP + 15% PP). During the preparation of the cementitious pastes, the workability was determined through the consistency index according to ASTM C1437 [28]. The water was added to the system until the consistency index reached 260 ± 10 mm.

2.2. Methods

2.2.1. Chemical, Thermal, and Mineralogical Characterization

The chemical composition was determined by X-ray fluorescence (EDX) in EDX 720 given by Shimadzu. The mineralogical characterization was performed by X-ray diffraction (XRD) in XRD6000 (Shimadzu) with radiation Kα-Cu (40 kV/30 mA), a goniometer speed of 2 min^{-1}, and a step of 0.02°. The XRD data analysis was performed using the PDF database 02 of the International Center for Diffraction Data (ICDD®). The interpretation was performed by comparison with standards contained in PDF 02 (ICDD, 2003) [29,30].

SEM images were acquired from a cross section of mortar bars (10 mm × 10 mm). A marble saw with a diamond disk was used to cut the samples. Before SEM images, a thin film (~50 nm) of gold was deposited at the cross section of the mortar bars. Such experiments were performed with the SANYU electron model SC-701, in which the experimental conditions were 2 min and 10 mA.

2.2.2. Resistance to Alkali–Aggregate Reaction

The resistance to the alkali–aggregate reaction was evaluated from linear expansion measurements performed on specimens with 25 mm × 25 mm × 285 mm dimensions, as determined by ASTM C1260 [31]. The samples were cured for 24 h in a humid chamber with 100% relative humidity and then immersed in water at a temperature of 80 ± 2 °C for 24 h. After water immersion, the samples were measured with a digital caliper (initial reading) and then transferred to the 1 N NaOH, where they were kept at 80 ± 2 °C for 28 days. The linear expansion measurements were obtained at intervals equal to 1, 7, 14, 16, and 28 days. After 28 days of curing, the samples were characterized by X-ray diffraction (XRD), scanning electron microscopy (SEM), and thermal gravimetric analysis (TGA).

The compressive strength (CS) was determined on cylindrical specimens (50 mm × 100 mm), according to ASTM C39/C39M [32]. The strengths were determined in two steps: first for the specimens immersed after 28 days in distilled water at 80 °C ± 2 °C and second for samples immersed in 1 N NaOH solution for 28 days at the same temperature. The results represented the average of three specimens and were determined in a universal mechanical testing machine (SHIMADZU AG-IS). The charge speed was 0.25 ± 0.05 MPa/s.

3. Results and Discussion

The PP and CP tailings presented high contents of SiO_2 and Al_2O_3 (70% and 15%, respectively). Such SiO_2 and Al_2O_3 contents plus values of alkaline equivalents (Na_2O_{eq}) greater than 5% can trigger characteristic expansions of alkali–aggregate reactions (Table 2). The amorphous alkaline gel is the main product of such reactions and has high water absorption, which can cause expansion and consequent cracking of the cementitious materials [33–35]. Table 2 presents the chemical composition of the PP and CP tailings used in this research. The PP tailings were used as filler to improve the packing of the system because they presented a maximum diameter (D_{max}) and fineness module (M_F) equal to 0.6 mm and 1.27, respectively (Figure 1). The D_{max} value measured from the CP tailings was 1.2 mm, and a fineness module was ~3.2; thus, it was used as the aggregate.

Table 2. Chemical composition of the perlite tailings investigated in this work.

Material	Oxides (%)									
	CaO	SiO_2	MgO	Al_2O_3	Fe_2O_3	SO_3	K_2O	Na_2O	[a] Na_2O_{eq}	[b] LOI
PP	1.5	73.0	0.6	15.3	1.1	0.1	3.3	3.0	5.2	2.1
CP	1.2	71.8	0.6	15.4	0.9	0.1	3.2	3.5	5.6	3.3

[a] Na_2O_{eq} = %Na_2O + 0.658 %K_2O; [b] LOI: loss on ignition.

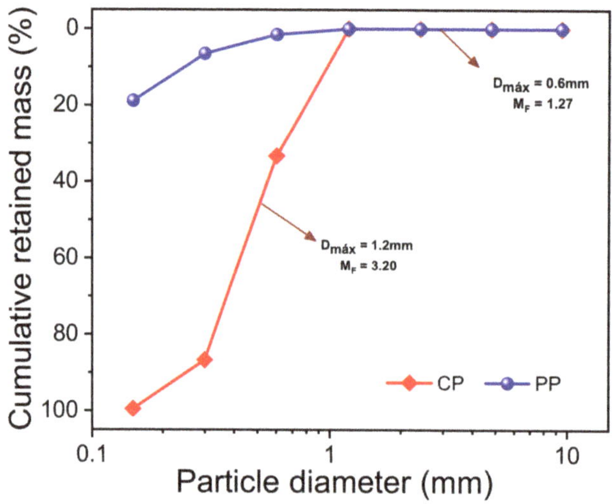

Figure 1. Particle size distribution of the PP and CP tailings.

Figure 2a,b show the linear expansion values measured from samples obtained from of the traces 1:3 (cement: 100% CP); 1:3 (cement: 85% CP and 15% PP); 1:1:6 (cement: lime: 100% CP), and 1:1:6 (cement: lime: 85% CP and 15% PP) immersed in a solution of 1 N NaOH for 28 days. For all samples, the linear expansion values measured at 16 days were above those previously seen in ASTM C 1260 [31], i.e., higher than 0.10%. The 1:3 (cement: 100% CP) and 1:3 (cement: 85% CP and 15% PP) samples presented linear expansions of 0.62% and 0.58%, while the 1:1:6 (cement: lime: 100% CP) and 1:1:6 (cement: lime: 85% CP and 15% PP) samples presented linear expansions of 0.86% and 0.66%, respectively. These results indicate that traces containing only the CP aggregate presented higher linear expansion than those containing CP and PP tailings. There is a high probability that this linear expansion above the expected is related to the formation of amorphous alkaline gel synthesized from the alkali–silicate reactions, induced by the high contents of the SiO_2, Al_2O_3, and alkaline oxides (Na_2O and K_2O) detected in the CP and PP tailings.

Still, in Figure 2a,b, the relatively smaller linear expansion measured at samples contained 15% of PP tailings can be attributed to the filler effect. As PP tailings are a powdery material (see Figure 1), its addition to the trace amplifies the packing of the cementitious matrix, decreasing the permeability of the system and the probability of forming the amorphous alkaline gel from the solutions contained in the pores [36].

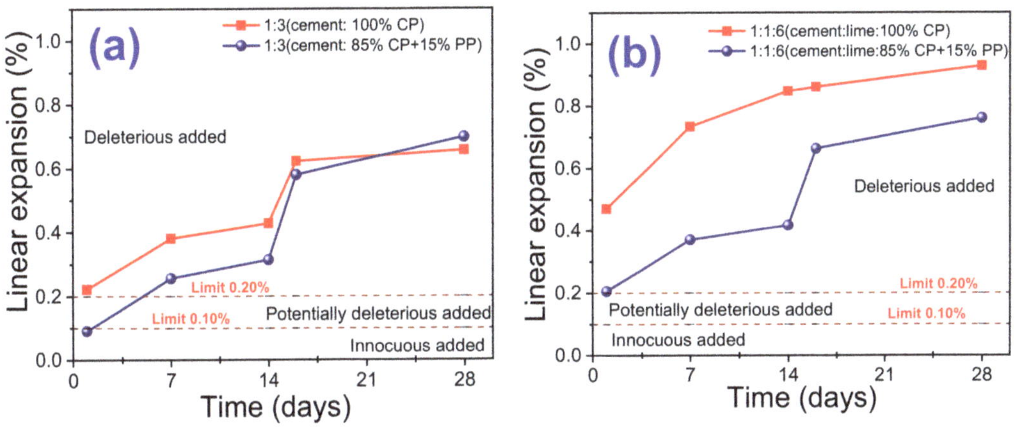

Figure 2. Linear expansions measured from the samples immersed in an aqueous solution of 1 N NaOH for 28 days: (**a**) traces 1:3 (cement: 100% CP) and 1:3 (cement: 85% CP +15% PP); (**b**) traces 1:1:6 (cement: lime: 100% CP) and 1:1:6 (cement: lime: 85% CP + 15% PP).

Figure 3 presents diffractograms measured from the perlite tailings (PP and CP) and samples after immersion in 1 N NaOH solution. The anorthoclase (JCPDS 09-0478) and quartz (JCPDS 46-1045) were the phases identified in the perlite tailings. The identification of aluminosilicate sodium and potassium phase (anorthoclase) favors the formation of the amorphous alkaline gel. It is known that such a gel is the main product of the alkali–aggregate reactions (RAAs). This is because, during immersion in an aggressive environment (1 N NaOH), the OH^- contained in pore solution comes into contact with the anorthoclase phase, make it soluble, and forms an amorphous alkaline gel. The amorphous band detected in the diffractograms of samples immersed in 1 N NaOH (Figure 3) is indicative of the formation of gels from the AAR [37]. Such a behavior is similar to that observed for silicates glasses, i.e., an amorphous band between 27° and 31° (2θ) with a strong reflection in 28° [38,39]. Other phases rich in sodium and potassium were detected in samples immersed in 1 N NaOH, i.e., calcium silicate (JCPDF 29-0368), sodium oxide (JCPDF 06-0500), sodium iron sulfate (JCPDF 27-0718), P-portlandite (JCPDF 04-0733), silicon oxide phases (JCPDF 27-0605), iron sulfide (JCPDF 15-0037), and anorthoclase (JCPDS 09-0478). The SEM images inserted in Figure 3 ratified the formation of the gel mentioned above. The areas marked with ellipses indicate the amorphous silica gel formed inside the pores of the mortars investigated in this study. Similar results were reported by Munhoz et al. [26].

Table 3 presents the compressive strength (CS) values of the mortars after immersion in 1 N NaOH and distilled water solutions. As expected, the samples immersed in 1 N NaOH solutions presented lower CS values than those immersed in distilled water. The 1:1:6 (cement:lime:100% CP) sample was the exception because CS values measured from them were equal (considered the standard deviation) in both experimental conditions. This behavior agrees with the linear expansion, XRD, and MEV results that confirmed the presence of amorphous alkaline gels in the samples kept in an aggressive environment. Additionally, in both experimental conditions investigated, the sample contends PP tailings presented higher CS values than samples without them. As mentioned before, this behavior occurred because the PP tailings worked as a filler material, causing a decrease in its porosity and consequently better packaging of the trace. Although the mortars studied in this work present high linear expansion values (see Figure 2), that is, above those established by ASTM C 1260, the compressive strength results indicate that such mortars can be used for masonry according to ACI 530.1-02/ASCE 6-02/TMS 602-02 [40].

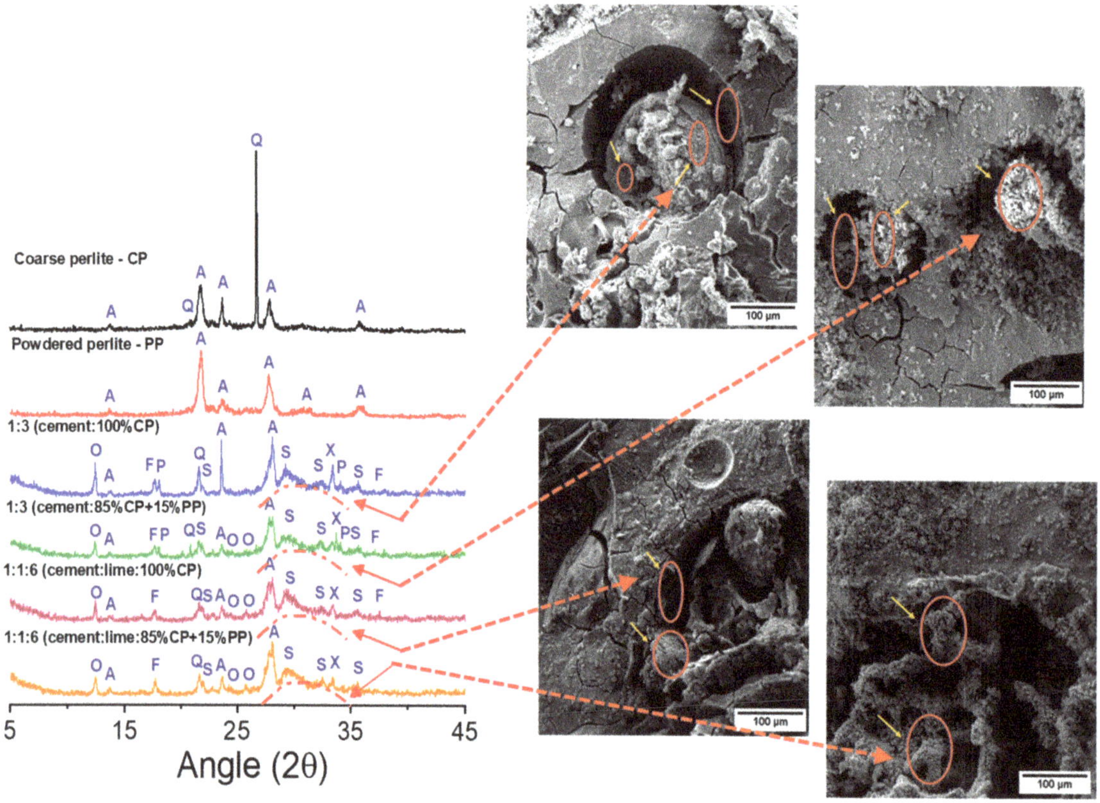

Figure 3. Diffractograms measured from as-received PP and CP tailings and mortars after the attack in 1 N NaOH solution. A—Anorthoclase; Q—Silicon oxide; S—Calcium silicate; X—Sodium oxide; F—Iron sulfide; O—Sodium iron sulfate; and P—Portlandite. SEM images were obtained from the mortars after attack in 1 N NaOH solution.

Table 3. Compressive strength of mortars after 28 day of immersion in 1 N NaOH and lime water.

Trace	Compression Strength (MPa) After Immersion in 1 N NaOH	Standard Deviation	After Immersion in Distilled Water	Standard Deviation
1:3 (cement:100% CP)	9.69	±0.7	13.04	±1.52
1:3 (cement:85% CP + 15% PP)	12.12	±1.1	17.07	±1.13
1:1:6 (cement:lime:100% CP)	5.80	±0.30	5.70	±0.13
1:1:6 (cement:lime:85% CP +15% PP)	8.32	±0.51	11.76	±1.34

4. Conclusions

Powdered and coarse perlite tailings were incorporated into mortars containing cement CP V-ARI and calcite lime. All linear expansion values measured from samples immersed in the 1 N NaOH solution were above those established by ASTM C 1260, i.e., >0.10%. This behavior was credited to the amorphous alkaline gel, which is the product of the alkali–aggregate reaction. This gel absorbs water, expands, and cracks the cementitious matrix. Indeed, the high contents of SiO_2, Al_2O_3, and alkaline oxides (Na_2O and K_2O) detected in perlite tailings favored alkali–silicate reactions. The presence of the gel was confirmed by XRD and SEM experiments. The cracks caused by the swelling of the alkaline gel negatively influenced the compressive strength values, where samples immersed in 1 N NaOH solution had lower CS values than those immersed in distilled water. Despite the high linear expansion values, the measured CS values indicate that the samples in this work can still be used in masonry, according to ACI 530.1 -02/ASCE 6-02/TMS 602-02.

Author Contributions: R.E.d.O.N. and J.d.M.C.: experimental data, data curation, and formal analysis; G.d.A.N. and R.R.M.: conceptualization, supervision, funding acquisition, and project administration; A.M.R., F.P.d.C., and S.V.A.B.: formal analysis, writing—original draft preparation, and writing—review and editing. All authors have read and agreed to the published version of the manuscript.

Funding: This research was funded by Coordenação de Aperfeiçoamento de Pessoal de Nível Superior (CAPES), grant number 88887.597478/2021-00.

Institutional Review Board Statement: Not applicable.

Informed Consent Statement: Not applicable.

Data Availability Statement: Not applicable.

Conflicts of Interest: The authors declare no conflict of interest.

References

1. Rashad, A.M. A synopsis about perlite as building material—A best practice guide for Civil Engineer. *Constr. Build. Mater.* **2016**, *121*, 338–353. [CrossRef]
2. El Mir, A.; Nehme, S.G. Utilization of industrial waste perlite powder in self-compacting concrete. *J. Clean. Prod.* **2017**, *156*, 507–517. [CrossRef]
3. Çelikten, S.; Işıkdağ, B. Properties of geopolymer mortars derived from ground calcined perlite and NaOH solution. *Eur. J. Environ. Civ. Eng.* **2021**, 1–15. [CrossRef]
4. Tie, T.S.; Mo, K.H.; Alengaram, U.J.; Kaliyavaradhan, S.K.; Ling, T.-C. Study on the use of lightweight expanded perlite and vermiculite aggregates in blended cement mortars. *Eur. J. Environ. Civ. Eng.* **2020**, 1–20. [CrossRef]
5. Alghadi, A.M.; Tirkes, S.; Tayfun, U. Mechanical, thermo-mechanical and morphological characterization of ABS based composites loaded with perlite mineral. *Mater. Res. Express* **2019**, *7*, 015301. [CrossRef]
6. El Mir, A.; Nehme, S.G.; Assaad, J.J. Durability of self-consolidating concrete containing natural waste perlite powders. *Heliyon* **2020**, *6*, e03165. [CrossRef] [PubMed]
7. Kotwica, Ł.; Pichór, W.; Kapeluszna, E.; Różycka, A. Utilization of waste expanded perlite as new effective supplementary cementitious material. *J. Clean. Prod.* **2017**, *140*, 1344–1352. [CrossRef]
8. Kapeluszna, E.; Kotwica, Ł.; Malata, G.; Murzyn, P.; Nocuń-Wczelik, W. The effect of highly reactive pozzolanic material on the early hydration of alite–C3A–gypsum synthetic cement systems. *Constr. Build. Mater.* **2020**, *251*, 118879. [CrossRef]
9. Almeida, E.; Carreiro, M.; Rodrigues, A.; Ferreira, H.; Santana, L.; Menezes, R.; Neves, G. A new eco-friendly mass formulation based on industrial mining residues for the manufacture of ceramic tiles. *Ceram. Int.* **2021**, *47*, 11340–11348. [CrossRef]
10. De Figueirêdo, J.M.R.; Da Costa, F.P.; Fernandes, J.V.; Rodrigues, A.M.; Neves, G.D.A.; Menezes, R.R.; Santana, L.N.D.L. Development of scheelite tailings-based ceramic formulations with the potential to manufacture porcelain tiles, semi-stoneware and stoneware. *Materials* **2020**, *13*, 5122. [CrossRef] [PubMed]
11. Fernandes, J.; Guedes, D.; Da Costa, F.; Rodrigues, A.; Neves, G.; Menezes, R.; Santana, L. Sustainable ceramic materials manufactured from ceramic formulations containing quartzite and scheelite tailings. *Sustainability.* **2020**, *12*, 9417. [CrossRef]
12. Gao, H.; Liu, H.; Liao, L.; Mei, L.; Shuai, P.; Xi, Z.; Lv, G. A novel inorganic thermal insulation material utilizing perlite tailings. *Energy Build.* **2019**, *190*, 25–33. [CrossRef]
13. Kayacı, K. The use of perlite as flux in the production of porcelain stoneware tiles. *Bol. Soc. Esp. Ceram. Vidr.* **2020**, *59*, 4–11. [CrossRef]
14. Ramezanianpour, A.; Karein, S.M.M.; Vosoughi, P.; Pilvar, A.; Isapour, S.; Moodi, F. Effects of calcined perlite powder as a SCM on the strength and permeability of concrete. *Constr. Build. Mater.* **2014**, *66*, 222–228. [CrossRef]
15. Polat, R.; Demirboğa, R.; Khushefati, W.H. Effects of nano and micro size of CaO and MgO, nano-clay and expanded perlite aggregate on the autogenous shrinkage of mortar. *Constr. Build. Mater.* **2015**, *81*, 268–275. [CrossRef]
16. Oktay, H.; Yumrutaş, R.; Akpolat, A. Mechanical and thermophysical properties of lightweight aggregate concretes. *Constr. Build. Mater.* **2015**, *96*, 217–225. [CrossRef]
17. Kramar, D.; Bindiganavile, V. Impact response of lightweight mortars containing expanded perlite. *Cem. Concr. Compos.* **2013**, *37*, 205–214. [CrossRef]
18. Erdoğan, S.T.; Sağlık, A. Ünsal Early-age activation of cement pastes and mortars containing ground perlite as a pozzolan. *Cem. Concr. Compos.* **2013**, *38*, 29–39. [CrossRef]
19. Guenanou, F.; Khelafi, H.; Aattache, A. Behavior of perlite-based mortars on physicochemical characteristics, mechanical and carbonation: Case of perlite of Hammam Boughrara. *J. Build. Eng.* **2019**, *24*, 100734. [CrossRef]
20. Różycka, A.; Pichór, W. Effect of perlite waste addition on the properties of autoclaved aerated concrete. *Constr. Build. Mater.* **2016**, *120*, 65–71. [CrossRef]

21. Angulo-Ramírez, D.E.; Mejía de Gutiérrez, R.; Medeiros, M. Alkali-activated Portland blast furnace slag cement mortars: Performance to alkali-aggregate reaction. *Constr. Build. Mater.* **2018**, *179*, 49–56. [CrossRef]
22. Rutkauskas, A.; Nagrockienė, D.; Skripkiūnas, G.; Barkauskas, K. Determining reactivity level of granite aggregate for con-crete. *Constr. Sci.* **2017**, *20*, 33–37. [CrossRef]
23. Ribeiro, D.V.; Silva, A.M.S.; Labrincha, J.A.; Morelli, M.R. Study of alkali-silica reactions associated with the use of red mud in plastering mortars. *Cerâmica* **2012**, *58*, 90–98. [CrossRef]
24. Rashidian-Dezfouli, H.; Rangaraju, P.R. Study on the effect of selected parameters on the alkali-silica reaction of aggregate in ground glass fiber and fly ash-based geopolymer mortars. *Constr. Build. Mater.* **2021**, *271*, 121549. [CrossRef]
25. Barros, S.V.A.; Marciano, J.E.A.; Ferreira, H.C.; Menezes, R.R.; Neves, G.A. Addition of quartzite residues on mortars: Analysis of the alkali aggregate reaction and the mechanical behaviour. *Constr. Build. Mater.* **2016**, *118*, 344–351. [CrossRef]
26. Munhoz, G.S.; Dobrovolski, M.E.G.; Pereira, E.; Medeiros-Junior, R.A. Effect of improved autogenous mortar self-healing in the alkali-aggregate reaction. *Cem. Concr. Compos.* **2020**, *117*, 103905. [CrossRef]
27. Neto, R.E.d.O.; Cartaxo, J.d.M.; Rodrigues, A.M.; Barros, S.V.A.; Costa, F.P.d.; Neves, G.d.A.; Menezes, R.R. New sustainable mortar compositions containing perlite residues. *Clean Technol. Environ. Policy* **2021**. submitted.
28. ASTM. *ASTM C1437-20. Standard Test Method for Flow of Hydraulic Cement Mortar*; ASTM International: West Conshohocken, PA, USA, 2020. [CrossRef]
29. Da Costa, F.P.; Morais, C.R.D.S.; Pinto, H.C.; Rodrigues, A.M. Microstructure and physico-mechanical properties of Al2O3-doped sustainable glass-ceramic foams. *Mater. Chem. Phys.* **2020**, *256*, 123612. [CrossRef]
30. Da Costa, F.P.; Morais, C.R.D.S.; Rodrigues, A.M. Sustainable glass-ceramic foams manufactured from waste glass bottles and bentonite. *Ceram. Int.* **2020**, *46*, 17957–17961. [CrossRef]
31. ASTM. *ASTM C1260-21 Standard Test Method for Potential Alkaline Reactivity of Aggregates (Mortar-Bar Method)*; ASTM International: West Conshohocken, PA, USA, 2021. [CrossRef]
32. ASTM. *ASTM C39/C39M-21. Standard Test Method for Compressive Strength of Cylindrical Concrete Specimens*; ASTM International: West Conshohocken, PA, USA, 2021. [CrossRef]
33. Xue, L.; Zhang, Z.; Wang, H. Hydration mechanisms and durability of hybrid alkaline cements (HACs): A review. *Constr. Build. Mater.* **2021**, *266*, 121039. [CrossRef]
34. Moreira, K.M.D.V.; Oliveira, P.V.G.; De Deus, Ê.P.; Cabral, A.E.B. Alkali-silica reaction: Understanding the phenomenon. *J. Build. Pathol. Rehabil.* **2021**, *6*, 1–10. [CrossRef]
35. Tiecher, F.; Florindo, R.N.; Vieira, G.L.; Gomes, M.E.B.; Molin, D.C.C.D.; Lermen, R.T. Influence of the Quartz Deformation Structures for the Occurrence of the Alkali–Silica Reaction. *Materials* **2018**, *11*, 1692. [CrossRef] [PubMed]
36. Romano, R.C.D.O.; Torres, D.D.R.; Pileggi, R.G. Impact of aggregate grading and air-entrainment on the properties of fresh and hardened mortars. *Constr. Build. Mater.* **2015**, *82*, 219–226. [CrossRef]
37. Filho, J.H.; Gobbi, A.; Pereira, E.; Quarcioni, V.A.; De Medeiros, M.H.F. Atividade pozolânica de adições minerais para cimento Portland (Parte I): Índice de atividade pozolânica (IAP) com cal, difração de raios-X (DRX), termogravimetria (TG/DTG) e Chapelle modificado. *Matéria* **2017**, *22*. [CrossRef]
38. Moulton, B.J.A.; Rodrigues, A.M.; Pizani, P.S.; Sampaio, D.; Zanotto, E.D. A Raman investigation of the structural evolution of supercooled liquid barium disilicate during crystallization. *Int. J. Appl. Glas. Sci.* **2018**, *9*, 510–517. [CrossRef]
39. Moulton, B.J.A.; Rodrigues, A.M.; Sampaio, D.V.; Silva, L.D.; Cunha, T.R.; Zanotto, E.D.; Pizani, P.S. The origin of the unusual DSC peaks of supercooled barium disilicate liquid. *Cryst. Eng. Comm.* **2019**, *21*, 2768–2778. [CrossRef]
40. Colville, J.; Abboud, B.E.; Atkinson, R.H.; Bailey, W.G.; Beavers, S.R.; Beiner, R.J.; Brown, R.H.; Bryant, A.D.; Callahan, K.D.; Catani, M.J.; et al. *Specification for Masonry Structures (ACI 530.1-02/ASCE 6-02/TMS 602-02)*; Masonry Standards Joint Committee: Reston, VA, USA, 1995.

sustainability

Article

Firing Parameters Effect on the Physical and Mechanical Properties of Scheelite Tailings-Containing Ceramic Masses

Marcos Emmanuel Araújo Carreiro [1], Valmir José da Silva [1], Alisson Mendes Rodrigues [1,2,*], Ester Pires de Almeida Barbosa [1], Fabiana Pereira da Costa [1], Romualdo Rodrigues Menezes [1,2], Gelmires Araújo Neves [1,2] and Lisiane Navarro de Lima Santana [1,2]

1 Graduate Program in Materials Science and Engineering (PPG-CEMat), Federal University of Campina Grande, Campina Grande 58429-900, Brazil; marcos_eac@hotmail.com (M.E.A.C.); valmir_jspb@yahoo.com.br (V.J.d.S.); esther.almeida@ufcg.edu.br (E.P.d.A.B.); fabiana.costa@estudante.ufcg.edu.br (F.P.d.C.); romualdo.menezes@ufcg.edu.br (R.R.M.); gelmires.neves@ufcg.edu.br (G.A.N.); lisiane.navarro@ufcg.edu.br (L.N.d.L.S.)

2 Laboratory of Materials Technology (LTM), Department of Materials Engineering, Federal University of Campina Grande (UFCG), Campina Grande 58429-900, Brazil

* Correspondence: alisson.mendes@professor.ufcg.edu.br

Abstract: The firing parameters in ceramic masses incorporated with 0, 5, and 10 wt% of scheelite tailings were investigated. The ceramic masses were characterized by X-ray fluorescence, granulometric, mineralogical analysis, and Atterberg limits determination. The samples were obtained by uniaxial pressing (20 MPa), sintered at different temperatures (800, 900, and 1000 °C), and heating rates (5, 10, 15, and 20 $°C \cdot min^{-1}$). Physical and mechanical tests (water absorption, apparent porosity, and flexural strength) and mineralogical tests were accomplished from the sintered samples. Natural aging tests were also carried out to assess carbonation resistance. For this, some samples were kept in an internal environment (inside the laboratory) for 3 months. The results showed a high content of calcium oxide in the scheelite tailings and a reduction in the plasticity index of the ceramic masses with the tailings addition. The best results were observed for the ceramic mass with 5% tailings. The best results were observed regarding the firing parameters for the temperature equal to 1000 °C, increasing the heating rate to 10 $°C \cdot min^{-1}$ without compromising the material properties. The samples kept in an internal environment for 3 months showed a loss of physical and mechanical properties. Such behavior probably occurred due to the onset of the carbonation phenomenon.

Keywords: scheelite tailings; sustainable ceramic mass; red ceramic; firing parameters

Citation: Carreiro, M.E.A.; da Silva, V.J.; Rodrigues, A.M.; Barbosa, E.P.d.A.; da Costa, F.P.; Menezes, R.R.; Neves, G.A.; Santana, L.N.d.L. Firing Parameters Effect on the Physical and Mechanical Properties of Scheelite Tailings-Containing Ceramic Masses. *Sustainability* **2022**, *14*, 333. https://doi.org/10.3390/su14010333

Academic Editor: Chongchong Qi

Received: 17 October 2021
Accepted: 26 November 2021
Published: 29 December 2021

Publisher's Note: MDPI stays neutral with regard to jurisdictional claims in published maps and institutional affiliations.

1. Introduction

The increase in the global population results in an unrestrained rise in waste generation (industrial, mining, domestic, hospital, etc.), which often does not have the correct disposal, causing severe social and environmental problems. The massive tailings amount generated in mining activities has stimulated research interest to use these tailings to develop friendly construction materials [1–4]. Thus, the construction industry has been the target of research due to its potential tailings reuse [5–7]. The construction industry has been considered the largest consumer of waste, generating sustainability, conserving natural resources, and mitigating the adverse effects of tailings on the environment.

On the other hand, the tailings incorporation into the ceramics industry makes it possible to reduce the extraction of natural raw materials, avoid degrading native areas, apply correct disposal and recycling of waste, and generate a product higher-value-added [8]. Traditional ceramic products, such as roof tiles, blocks, and ceramic tiles, allow incorporating different tailings types due to their diversified composition. In recent years, some research has been carried out to incorporate the most varied types of tailings in ceramic masses destined for traditional ceramic pieces production [9–14]. In this sense, several mining tailings, such as kaolin waste [9,10], waste from phosphate mines [11], iron ore tailings [12], red mud from the Bayer process [13,14], tailings from silver mines [15], quartzite [5,16,17], and scheelite [5,6] have been successfully reused in the production of sustainable ceramic materials.

Scheelite ($CaWO_4$) is a typical tungstate mineral, which is used to produce tungsten, which is usually associated with other minerals that contain calcium, such as calcite ($CaCO_3$), fluorite (CaF_2), and fluorapatite [$Ca_{10}(PO_4)6F_2$] [1,18,19]. In beneficiation, the ore goes through the screening, crushing, and grinding processes, placed in jigues that feed the vibrating tables which separate the concentrate from the tailings. In this last stage, the scheelite tailings are generated, which has no use for the industry and is improperly deposited.

An important issue for the use of waste as an alternative raw material in the ceramic field is the knowledge of its chemical components, in addition to the mineralogical and physical characteristics [20]. Products obtained with clay-based compositions can be reinforced using mixtures of clay and calcite to form micro-composites, such as anorthite grains ($CaO{\cdot}Al_2O_3{\cdot}2SiO_2$) in the silico-aluminous matrix, which favors a significant increase of resistance [21]. Therefore, the addition of tailings rich in calcium oxide, such as scheelite tailings, may favor the formation of anorthite.

It is well known that the increase in the firing temperature in clay-based ceramics produces a series of reactions and transformations that lead to the formation of new phases and the disappearance of others [22]. Therefore, it is necessary to analyze the quality of the ceramic masses, monitor, and understand the mineralogical transformations that occur during the firing process. The insertion of rapid firing in the ceramic material firing stage has been studied [23,24]; however, much still needs to be investigated, especially in masses containing tailings.

On the other hand, calcium silicate phases, such as gehlenite and anorthite, are potential sources of CaO and can favor carbonation. Carbonation is the process in which the CO_2 present in the atmosphere reacts with calcium hydroxide forming calcium carbonate that precipitates inside the microstructure, which can change its porosity [25,26]. Carbonation occurs mainly in materials rich in calcium oxide and can deteriorate the material with property loss [22]. Most research on carbonation is carried out with a focus on concrete and cementitious materials [27,28]. However, the carbonation process also occurs in ceramic construction materials, and studies aimed at these materials still need to be better explored since carbonation can affect its durability [29].

Although the use of scheelite tailings in the production of ceramic materials has already been reported in the literature, the influence of these tailings on the properties of traditional ceramic products, especially regarding carbonation resistance, still needs to be studied. Therefore, the objective of this work was to analyze the physical and mechanical properties and the carbonation resistance of samples obtained from the ceramic masses (used in the manufacture of ceramic blocks) containing 0, 5, and 10% scheelite tailings. The influence of different firing parameters (temperature (800, 900, and 1000 °C) and heating rate (5, 10, 15, and 20 $°C{\cdot}min^{-1}$) on physical and mechanical properties (water absorption, apparent porosity, and strength flexural) was investigated. The carbonation resistance was evaluated in the specimens after 3 months of exposure to natural environmental conditions.

2. Materials and Methods

2.1. *Raw Materials*

Scheelite tailings were obtained from a tungsten mining company (Brejuí mine) located at Currais Novos city, Rio Grande do Norte state, Brazil. Three types of scheelite tailings from different mining stages were used to obtain the samples. Such tailings were denominated ST1, ST2, and ST3. The ceramic mass used as standard (MR) was donated by a red ceramic company located at Soledade city, Paraíba state, Brazil. The as-received scheelite tailings (ST1, ST2, and ST3) are shown in Figure 1a–c.

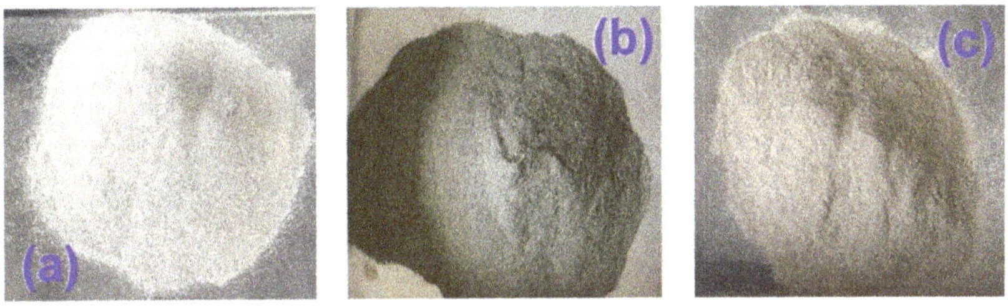

Figure 1. Scheelite tailings investigated (**a**) ST1, (**b**) ST2, and (**c**) ST3.

2.2. *Incorporation of Scheelite Tailings in the Ceramic Mass and Preparation of Specimens*

The scheelite tailings (ST1, ST2, and ST3) were added to MR at different contents (0 wt%, 5 wt%, and 10 wt%). The choice of incorporation contents of 5 wt% and 10 wt% of scheelite tailings were based on previous studies developed by Machado et al. [30]. The new ceramic masses were inserted into a porcelain jar containing alumina balls and dry homogenized (7 wt% humidity) for 24 h in a ball mill. The consistency of the masses was also evaluated by liquidity and plasticity limits using the Casagrande method [31]. Table 1 shows the nominal composition of the prepared ceramic masses.

Table 1. Nominal composition of the ceramic masses incorporated with scheelite tailings.

Ceramic Masses	Addition Content (%wt)	Scheelite Tailings Type
MR	0	-
M5S1	5	ST1
M5S2	5	ST2
M5S3	5	ST3
M10S1	10	ST1
M10S2	10	ST2
M10S3	10	ST3

The new ceramic masses were uniaxially pressed (20 MPa) in a hydraulic press (Servitech, model CT-335, Tubarão, Brazil) to obtain samples with dimensions of 50 mm × 15 mm × 5 mm. The samples were dried (110 °C for 24 h) and fired in an electric oven (Flyever Equipment, model FE50RP, São Carlos, Brazil) at different temperatures (800, 900, and 1000 °C) and heating rates (5, 10, 15 and 20 $°C·min^{-1}$). An isothermal treatment of 30 min was accomplished at each firing temperature and then the samples were cooled for room temperature at 10 $°C·min^{-1}$.

2.3. *Characterizations*

The chemical composition of scheelite tailings and of ceramic masses were determined by X-ray fluorescence (XRF) (Shimadzu, model EDX 720, Kyoto, Japan). The analysis was carried out in triplicate and performed under vacuum, with a collimator of 10 mm, a Rh X-ray tube, and an EDS detector. The mineralogical composition was determined by X-ray diffraction (XRD) (Shimadzu, model XRD 6000, Kyoto, Japan) with Cu-Kα radiation (40 Kv/30 mA); goniometer speed $2°·min^{-1}$; 0.02° step; and 2θ scanning from 2° to 60°. The granulometric analysis (GA) of ceramic masses was determined by laser diffraction (Cilas, model 1064, Orléans, France). All raw materials were sieved (74 μm) before XRF, XRD, and GA analyses.

Water absorption (WA), apparent porosity (AP), and 3-point flexural strength (FS) tests were performed on the specimens after firing. WA and AP were measured by Archimedes method. In summary, the samples were immersed in distilled water for 24 h to obtain the wet weight (W_w) and the immersed weight (W_i). Before immersion in distilled water, sintered samples were weighed to obtain their dry weight (W_d). Equations (1) and (2) were used to calculate WA and AP, respectively.

The flexural rupture module experiments were carried with the aid of a universal mechanical testing machine (Shimadzu, model Autograph AG-X 50 kN, Kyoto, Japan), with 5 KN load cells, the distance between the support points of 30 mm, and a test speed of 0.5 mm·min^{-1}. The results of physical and mechanical tests were obtained from an average of 10 samples. The mineralogical phases formed after firing was analyzed by X-ray diffraction (XRD).

$$\mathrm{WA}(\%) = \frac{\mathrm{W_w} - \mathrm{W_d}}{\mathrm{W_d}} \times 100, \quad (1)$$

$$\mathrm{AP}(\%) = \frac{\mathrm{W_w} - \mathrm{W_d}}{\mathrm{W_w} - \mathrm{W_i}} \times 100, \quad (2)$$

where W_w, W_d, and W_i are the weights of the specimens, wet, dry, and immersed, respectively.

2.4. Carbonation Resistance

The carbonation resistance evaluation was carried out on fired samples at 1000 °C at a heating rate of 5 °C·min^{-1}, which were submitted to the natural aging test in an internal environment. The samples were kept protected from external climatic actions (inside the laboratory, at room temperature, and under air humidity ~74%) for 3 months. After this period, the samples were evaluated to physical and mechanical (WA, AP, and FS), mineralogical (XRD), and fracture surface analysis using scanning electron microscopy (SEM).

3. Results and Discussion

3.1. Chemical and Mineralogical Composition of the Scheelite Tailings and Ceramic Masses

Table 2 presents the chemical composition of the scheelite tailings used in this research and the ceramic masses without and with scheelite tailings. The ceramic mass without scheelite tailings was denominated MR, while with tailings were denominated M5S1, M5S2, and M5S3 (with 5 wt%) and M10S1, M10S2, and M10S3 (with 10 wt%). The calcium oxide (CaO) was a major component detected in the scheelite tailings, which contents ranging from 37.6 to 40.2%, and is related to the presence of calcite or dolomite [5]. Additionally, SiO_2 (18.1 to 22.9%), Al_2O_3 (7.6 to 10.8%), and Fe_2O_3 (7.2 to 9.8%) contents were detected in the scheelite tailings. The ceramic masses presented a typical composition for the manufacture of red ceramic products. A high Fe_2O_3 content (8.2–9.0 wt%) is responsible for the reddish color of the pieces after firing. SiO_2 and Al_2O_3 were the major oxides detected. These oxides have origin from the clay minerals, feldspar, and free silica presented in the scheelite tailings. In the ceramic masses containing scheelite tailings (M5S1, M5S2, M5S3, M10S1, M10S2, and M10S3), an increase in the CaO content in relation to the MR mass (without tailings) is observed due to the higher concentration of this oxide in the tailings. In general, ceramic masses containing tailings have a high content of fluxing oxides (K_2O, CaO, and MgO), responsible for forming glass phases, reducing the firing temperature, and improvingthe physical and mechanical properties [7]. The loss on ignition (LOI) measured from the scheelite tailings is associated with free water loss and carbonate decompositions. In the ceramic masses, LOI is due to free water loss, clay mineral dehydroxylation, organic matter, and carbonate decompositions. All ceramic masses containing scheelite tailings (M5S1, M5S2, M5S3, M10S1, M10S2, and M10S3) presented a higher percentage of LOI compared to the ceramic mass without the tailings (MR). Such behavior is probably due to the increase in carbonate content.

Figure 2a shows XRD patterns measured from standard ceramic mass (MR), i.e., without the scheelite tailings. Quartz (JCPDS 46-1045), feldspar (JCPDS 84-0710), and kaolinite (JCPDS 79-1570) were the main crystalline phases identified in MR. On the other hand, mica (JCPDS 83-1808) and calcite (JCPDS 72-1937) phases were identified to a lesser extent. The XRD patterns measured from scheelite tailings (ST1, ST2, and ST3) are shown in Figure 2b. In general, all scheelite tailings presented the following mineralogical phases: calcite (JCPDS 72-1937), quartz (JCPDS 46-1045), and diopside (JCPDS 89-0837). Diopside usually occurs as an accessory mineral and is associated with calcite. Peaks referring to dolomite (JCPDS 89-5862) were identified at ST2 and ST3. These peaks may be associated with a higher MgO (3.3%) and CaO (38.5 and 40.2%) content detected in these tailings, compared to ST1, which presented 2.7% MgO and 37.6% CaO. In general, the minerals detected in scheelite tailings are generally present in the masses used in traditional ceramics, which favor the incorporation of these tailings in this sector.

Table 2. Chemical composition of scheelite tailings (ST1, ST2, and ST3) used in this study and of ceramic masses without (MR) and with scheelite tailings (5 and 10%). The results were determined by X-ray fluorescence.

Raw Materials/Samples	Oxides (%wt)										
	SiO_2	Al_2O_3	Fe_2O_3	K_2O	TiO_2	CaO	MgO	WO_3	Others	LOI *	SiO_2/Al_2O_3
ST1	18.1	8.7	7.2	0.4	0.6	37.6	2.7	0.4	1.3	23.0	-
ST2	21.8	7.6	9.8	0.6	0.7	38.5	3.3	0.2	1.8	15.7	-
ST3	22.9	10.8	9.6	0.6	0.4	40.2	3.3	0.7	1.7	9.8	-
MR	47.5	20.9	8.9	3.0	1.2	2.1	2.6	-	0.5	13.3	2.27
M5S1	45.0	19.8	8.2	2.9	1.2	3.5	2.5	-	0.6	16.3	2.27
M5S2	43.7	19.4	9.0	2.7	1.0	4.8	2.5	-	1.0	15.9	2.25
M5S3	45.8	20.0	9.0	2.9	1.2	3.0	2.5	-	0.6	15.0	2.29
M10S1	43.5	19.4	8.5	2.8	1.1	3.7	2.5	-	0.8	17.7	2.24
M10S2	43.0	19.0	9.0	2.7	1.0	5.7	2.5	-	0.9	16.2	2.26
M10S3	42.9	18.9	8.7	2.6	1.1	7.2	2.4	-	0.6	15.6	2.27

* LOI—Loss on ignition measured after drying at 110 °C and firing at 1000 °C.

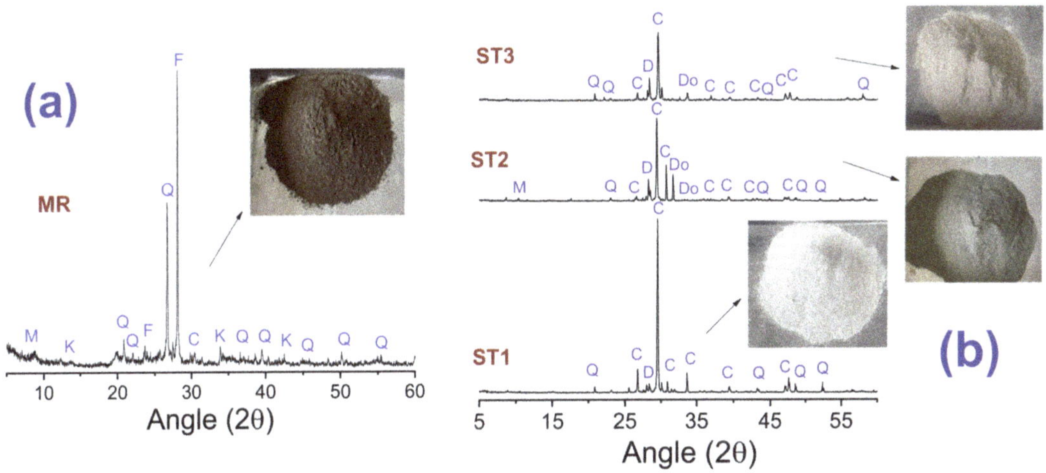

Figure 2. XRD patterns measured from ceramic mass (MR) (**a**) and of the scheelite tailings (ST1, ST2, and ST3) (**b**). M—Mica; K—Kaolinite; F—Feldspar; Q—Quartz; C—Calcite; D—Diopiside; Do—Dolomite.

3.2. Granulometry and Consistency of Ceramic Masses

Knowledge of the particle size distribution and consistency of the ceramic masses is extremely important before the molding step of the ceramic product. Through the results of granulometry and consistency values, it is possible to assess whether the raw materials used are suitable for the molding and manufacturing process of ceramic pieces. In addition, the variation in particle size distribution and plasticity of the ceramic masses can cause changes in their behavior during processing and significantly affect the properties of the final products.

The accumulated mass percentage for different particle size ranges and the average diameter of particles obtained from the ceramic masses are presented in Table 3. All ceramic masses showed similar particle size distributions, being approximate: 21% clay fraction (<2 μm), 57% silt fraction (between 2 and 20 μm), and 21% sand fraction (>20 μm). The average diameter of the ceramic masses was approximately 12 μm. In general, the ceramic masses have a wide granulometric distribution, which can favor the packing of the particles during compaction and the reactivity during firing temperature. It is worth mentioning that the reactivity between the particles favors the reactions for the formation of new crystalline phases based on calcium and/or magnesium, which significantly affects the physical and mechanical properties of the sintered parts [32,33].

Table 3. Accumulated mass for different particle size distribution ranges and average diameter of the ceramic masses.

Samples	Accumulated Mass (%)			Average Diameter (μm)
	Fine (x * < 2 μm)	Medium (2 μm < x < 20 μm)	Gross (x > 20 μm)	
MR	22.1	55.5	22.4	12.5
M5S1	20.4	58.8	20.8	12.2
M5S2	21.5	57.3	21.2	12.9
M5S3	20.9	56.2	22.9	12.6
M10S1	20.6	57.6	21.8	12.1
M10S2	20.5	57.6	21.9	12.6
M10S3	20.6	57.8	21.6	12.4

* x: particle size.

Figure 3 shows the Atterberg limits measured from the ceramic masses with and without scheelite tailings. In general, there is a plasticity index reduction with the incorporation of scheelite tailings. Such behavior is probably related to the non-plastic characteristic of the scheelite tailings. However, the plasticity and liquidity indices and plasticity limits are within the range of the literature for use in red ceramics [34]. Ceramic masses with a plasticity index between 7 and 15% are classified as moderately plastic and values above 15% as highly plastic. In this way, the ceramic masses studied in this work are highly plastic, making them viable for the red ceramics pieces production.

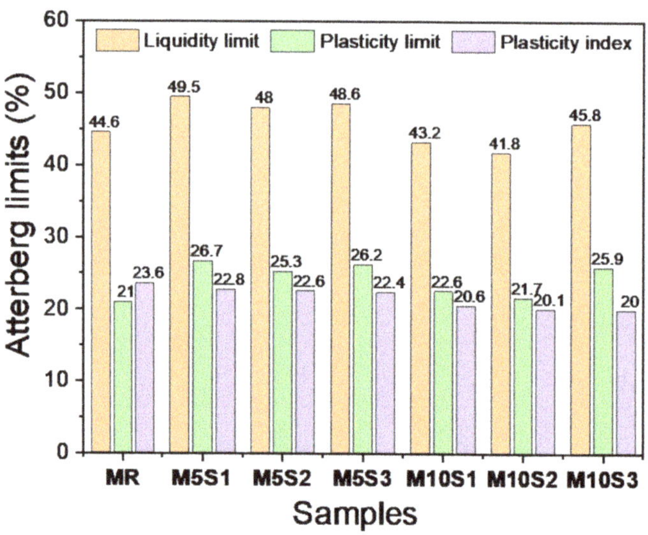

Figure 3. Atterberg limits (liquidity limit, plasticity limit, and plasticity index) of the ceramic masses investigated.

3.3. Mineralogical Phases of the Samples after Firing

Figure 4 shows the X-ray diffraction patterns of samples sintered at 800, 900, and 1000 °C, with a heating rate of 5 °C min^{-1}. The following crystalline phases were observed: mica ($KMg_3(Si_3Al)O_{10}(OH)_2$, JCPDS 83-1808), quartz (SiO_2, JCPDS 46-1045), feldspar (JCPDS 84-0710), anorthite ($CaAl_2Si_2O_8$, JCPDS 89-1472), and gehlenite ($Ca_2Al(AlSi)O_7$, JCPDS 35-0755). Peaks characteristic of the mica phase were detected only at 800 and 900 °C, where it was possible to observe a reduction in its intensity when the temperature increased from 800 °C to 900 °C. Feldspar is present in all compositions and temperatures; however, the peak intensity decreases as the temperature increases. The kaolinite phase was not identified in any investigated samples after sintering, showing that this phase detected in the ceramic mass before firing (see Figure 2a) disappeared due to dehydroxylation between 500 and 600 °C [35,36]. After dehydroxylation, the layered structure of kaolinite is destroyed, transforming into an amorphous phase called metakaolinite. At temperatures around 900 °C, metakaolinite reorganizes and forms other crystalline phases such as mullite, hematite, gehlenite, and anorthite. The formation of these phases depends on the amount of other minerals present in the material [26].

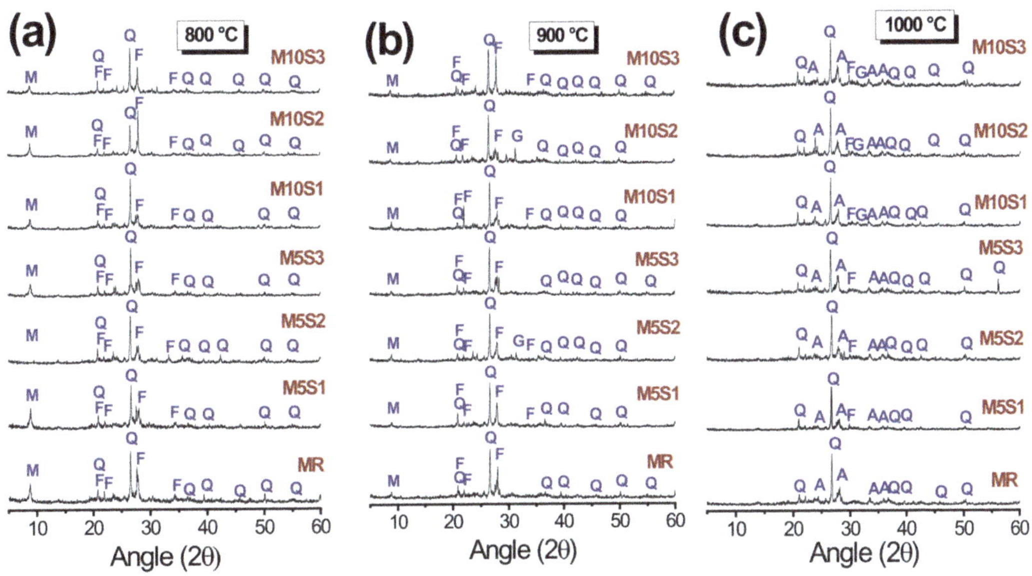

Figure 4. XRD patterns measured from samples sintered at 800 (**a**), 900 (**b**), and 1000 °C (**c**). All samples were fired at a heating rate of 5 °C·min^{-1}. M—Mica; F—Feldspar; Q—Quartz; A—Anorthite; G—Gehlenite.

In the scheelite tailings (ST1, ST2, and ST3), incorporated MR calcite ($CaCO_3$) was the main mineralogical phase identified (see Figure 2b). According to Baccour et al. [28], when the temperature increase from 850 to 950 °C, $CaCO_3$ decomposes ($CaO + CO_2$), and the formed CaO reacts with the amorphous phase (metakaolinite), forming crystalline phases, such as gehlenite and anorthite. It is worth mentioning that the first peaks related to the gehlenite phase occurred for the ceramic masses fired at 900 °C and 1000 °C to the anorthite. The appearance of the anorthite probably is related to the gehlenite into anorthite transformation. According to the literature, the presence of structural iron favors the metakaolinite to gehlenite transformation, which, later combines with aluminum, silicon (from metakaolinite), and quartz, form anorthite [21,37].

According to Jiang et al. [38], the gehlenite into anorthite transformation can be attributed to the high diffusion rate of Ca^{2+}. In anorthite, the Ca^{2+}:Si^{4+} ratio is 1:2, while for gehlenite is 2:1. The Ca^{2+} and O^{2-} ionic bond can be more easily destroyed because it is less stable than the Si-O covalent bond. As a result, at low temperatures, the Ca^{2+} diffusion rate was much faster than that of Si^{4+}, causing the appearance of gehlenite peaks at ~900 °C.

Clays compositions containing high CaO contents, fired at 950 and 1100 °C, were studied by González-Garcia et al. [39]. In such a study, the researchers associated the transformation of the gehlenite into anorthite with the presence of the illite, SiO_2, and $CaCO_3$. It was observed that the increase in temperature favored the gehlenite into anorthite transformation. The authors also highlighted the mullite absence in clays with significant $CaCO_3$ proportions. According to Ouahabi et al. [40], the CaO presence in the limestone clays prevents mullite formation. Quartz was detected in the ceramic masses before firing. No phase transformation was observed during the different heat treatments for this mineral, only reducing peak intensity at a firing temperature of 1000 °C.

3.4. Physical and Mechanical Properties of Sintered Samples

Table 4 presents water absorption (WA), apparent porosity (AP), and flexural strength (FS) values measured from samples fired at 800, 900, and 1000 °C using different heating rates (5, 10, 15, and 20 °C min^{-1}). For samples fired at 800 °C, an increase in the scheelite tailings contents resulted in a slight increase in the WA and AP values. Such behavior can be credited to carbonate decomposition. During the firing step, the calcite ($CaCO_3$) present in the scheelite tailings decomposes ($CaCO_3 \rightarrow CaO + CO_2$), and CO_2 formed increases the pressure inside the pores of the material, compromising the densification process [6]. Besides, for all heating rates used, it is observed that samples containing scheelite tailings (M5S1, M5S2, M5S3, M10S1, M10S2, and M10S3) tend to have higher FS values than the sample without tailings (MR), which is probably related to the interaction between the crystalline phases [21,41].

At heating rates above 10 °C min^{-1}, a reduction in mechanical strength was observed. Such behavior may be associated with the microcracks formation due to the faster cycle, especially at cooling, which occurs during the quartz polymorphic transformation ($\beta \rightarrow \alpha$). In general, the samples fired at 800 °C presented the potential to be used in the production of bricks because they showed water absorption values (range 15.5%–20.6%) within the maximum limit of 22% (ASTM C20-00 [42]) and flexural strength between 1–5 MPa. According to ASTM C67 [43], the minimum requirement for flexural strength of bricks is 0.65 MPa.

For samples sintered at 900 °C, the best condition (lowest WA and AP values and highest FS value) was obtained for M5S1 and M5S3 samples (5% scheelite tailings). These results were higher than those measured to MR samples (without tailings). Taking account of the crystalline phases formed during the firing step (at 900 °C) (Figure 4b), it is possible to observe a reduction in the intensity of the mica phase peaks and an increase in the intensity anorthite peaks. Such behavior can improve properties since the anorthite favors increased resistance and chemical stability of the material [21,41]. In general, the increase in firing temperature from 800 °C to 900 °C resulted in improved properties (WA and AP reduction and increase of FS), with the best results observed for samples heated to a rate of 5 °C min^{-1}. The FS values obtained were higher than those found by other researchers who worked in the same temperature range [44].

At 1000 °C and heating rates of 5 and 10 °C min^{-1}, the M5S1 sample showed FS values higher than the MR sample. This indicates that for this composition, the firing time can be reduced by increasing the rate from 5 to 10 °C min^{-1} without loss of mechanical properties. In general, the best properties were observed for samples fired at 1000 °C. This probably was due to the increase in the anorthite amount (Figure 4c), which contributes to the rise in the material's chemical resistance and stability and improves the physical properties [21,41,45].

Table 4. Physical and mechanical properties of the standard mass (MR) and ceramic masses with 5 and 10% scheelite tailings after sintering at 800, 900, and 1000 °C.

Rate (°C·min^{-1})	Samples	Physical and Mechanical Properties of the Masses								
		Water Absorption (%)				Apparent Porosity (%)			Flexural Strength (MPa)	
		800 °C	900 °C	1000 °C		800 °C	900 °C	1000 °C	800 °C	900 °C 1000 °C
	MR	17.7 ± 0.3	17.5 ± 0.2	13.2 ± 0.1	32.2 ± 0.4	31.7 ± 0.3	26.0 ± 0.2	2.0 ± 0.2	3.2 ± 0.3	6.4 ± 0.4
	M5S1	19.9 ± 0.3	16.0 ± 0.3	12.4 ± 0.3	35.1 ± 0.4	30.6 ± 0.6	24.5 ± 0.5	2.0 ± 0.4	5.6 ± 0.5	7.0 ± 0.8
	M5S2	17.6 ± 0.4	18.0 ± 0.2	13.9 ± 0.2	32.6 ± 0.6	32.9 ± 0.6	26.7 ± 0.3	3.6 ± 1.2	2.5 ± 0.2	5.5 ± 0.6
5	M5S3	15.5 ± 0.1	14.7 ± 0.2	11.9 ± 0.2	29.8 ± 0.2	27.9 ± 0.3	23.8 ± 0.3	5.0 ± 0.3	4.7 ± 0.4	6.7 ± 0.5
	M10S1	18.9 ± 0.3	16.0 ± 0.2	13.0 ± 0.2	34.5 ± 0.4	29.9 ± 0.3	25.3 ± 0.3	2.3 ± 0.5	4.1 ± 0.2	5.1 ± 0.3
	M10S2	18.8 ± 0.5	16.9 ± 0.3	13.1 ± 0.3	33.7 ± 0.4	31.4 ± 0.4	27.3 ± 0.4	3.5 ± 0.5	2.3 ± 0.2	4.1 ± 0.5
	M10S3	18.6 ± 0.4	14.6 ± 0.3	13.7 ± 0.4	34.1 ± 0.5	27.9 ± 0.4	26.8 ± 0.4	4.5 ± 0.2	4.0 ± 0.2	5.2 ± 0.3
	MR	17.2 ± 0.5	17.4 ± 0.3	13.4 ± 0.2	31.1 ± 0.5	31.6 ± 0.4	26.2 ± 0.3	1.4 ± 0.4	3.1 ± 0.2	6.4 ± 0.4
	M5S1	19.5 ± 0.4	16.3 ± 0.3	12.8 ± 0.4	34.5 ± 0.6	30.8 ± 0.5	25.1 ± 0.5	1.8 ± 0.2	5.0 ± 0.4	6.6 ± 0.4
	M5S2	17.5 ± 0.3	18.3 ± 0.2	14.4 ± 0.4	32.3 ± 0.5	33.2 ± 0.5	27.5 ± 0.6	3.6 ± 1.6	2.1 ± 0.1	4.5 ± 0.5
10	M5S3	15.6 ± 0.2	14.8 ± 0.1	12.4 ± 0.3	29.8 ± 0.3	28.0 ± 0.2	24.6 ± 0.5	4.0 ± 0.3	4.5 ± 0.4	5.4 ± 0.3
	M10S1	19.8 ± 0.4	16.4 ± 0.4	15.1 ± 0.3	35.5 ± 0.5	30.3 ± 0.5	28.1 ± 0.5	1.5 ± 0.2	3.2 ± 0.3	3.6 ± 0.4
	M10S2	19.5 ± 0.3	17.3 ± 0.3	15.3 ± 0.4	34.4 ± 0.4	32.1 ± 0.4	28.4 ± 0.6	3.4 ± 0.2	2.0 ± 0.1	3.6 ± 0.2
	M10S3	19.0 ± 0.2	14.6 ± 0.3	14.8 ± 0.2	34.5 ± 0.2	28.0 ± 0.4	27.8 ± 0.3	3.8 ± 0.3	3.6 ± 0.4	4.7 ± 0.4
	MR	17.5 ± 0.2	17.6 ± 0.1	13.2 ± 0.9	31.5 ± 0.3	31.9 ± 0.2	25.9 ± 1.3	1.0 ± 0.4	2.8 ± 0.2	6.7 ± 1.0
	M5S1	19.8 ± 0.2	16.5 ± 0.3	13.3 ± 0.3	34.8 ± 0.3	31.0 ± 0.4	25.9 ± 0.5	1.6 ± 0.2	4.5 ± 0.2	5.6 ± 0.5
	M5S2	17.5 ± 0.4	18.4 ± 0.2	15.1 ± 0.3	32.3 ± 0.5	33.3 ± 0.5	28.4 ± 0.4	3.0 ± 1.2	2.0 ± 0.2	4.1 ± 0.5
15	M5S3	15.8 ± 0.3	14.8 ± 0.2	13.1 ± 0.5	30.0 ± 0.4	28.0 ± 0.4	25.5 ± 0.7	3.7 ± 0.5	4.4 ± 0.3	4.7 ± 0.6
	M10S1	19.9 ± 0.2	17.3 ± 0.5	16.0 ± 0.4	35.5 ± 0.2	31.6 ± 0.7	29.2 ± 0.4	1.5 ± 0.3	2.9 ± 0.4	2.9 ± 0.2
	M10S2	19.5 ± 0.2	18.0 ± 0.3	15.7 ± 0.3	34.4 ± 0.3	32.9 ± 0.4	28.9 ± 0.4	2.9 ± 0.2	1.9 ± 0.2	3.6 ± 0.3
	M10S3	19.4 ± 0.5	15.1 ± 0.3	15.1 ± 0.4	35.0 ± 0.7	28.4 ± 0.5	28.2 ± 0.6	3.4 ± 0.2	3.1 ± 0.2	4.4 ± 0.2
	MR	17.7 ± 0.2	17.6 ± 0.2	13.9 ± 0.2	31.7 ± 0.3	31.8 ± 0.3	26.8 ± 0.3	1.0 ± 0.1	2.6 ± 0.2	5.7 ± 0.4
	M5S1	19.8 ± 0.3	16.6 ± 0.3	14.3 ± 0.5	34.9 ± 0.4	31.0 ± 0.5	27.2 ± 0.7	1.4 ± 0.3	4.3 ± 0.4	4.3 ± 0.8
	M5S2	18.2 ± 0.5	18.6 ± 0.2	15.8 ± 0.4	33.1 ± 0.6	33.5 ± 0.6	29.2 ± 0.5	2.3 ± 1.2	1.6 ± 0.2	3.4 ± 0.4
20	M5S3	16.4 ± 0.3	15.0 ± 0.2	13.9 ± 0.7	30.9 ± 0.4	28.2 ± 0.2	26.7 ± 1.0	2.6 ± 0.3	3.8 ± 0.3	3.5 ± 0.6
	M10S1	20.0 ± 0.1	18.8 ± 0.5	16.6 ± 0.5	36.5 ± 0.3	33.4 ± 0.6	30.0 ± 0.7	1.3 ± 0.3	2.1 ± 0.1	2.5 ± 0.3
	M10S2	19.8 ± 0.6	19.0 ± 0.3	15.7 ± 0.2	35.1 ± 0.3	33.9 ± 0.4	29.0 ± 0.2	2.1 ± 0.1	1.4 ± 0.2	3.4 ± 0.2
	M10S3	20.6 ± 0.8	15.4 ± 0.2	15.0 ± 0.3	36.1 ± 0.9	29.0 ± 0.3	28.0 ± 0.5	2.3 ± 0.1	2.9 ± 0.1	4.6 ± 0.3

Sintering is one of the main steps during processing a ceramic product; because it is at this stage, the consolidation of the part occurs. Factors such as firing temperature and heating rate can affect the sintering degree and, consequently, the physical and mechanical properties of the material [23]. Therefore, it is worth noting that the samples' quick-firing (i.e., higher heating rates) did not significantly compromise the mechanical properties. At temperatures of 800 and 900 °C, there was no considerable variation in the properties with the increase in the heating rate. This fact is relevant for the industry since rapid burning in the manufacture of ceramic parts reduces energy expenditure and saves companies.

Regarding the incorporation of the scheelite tailings, in general, the difference in the tailings type (ST1, ST2, and4 ST3) did not cause significant changes in physical and mechanical properties. This is because the three tailings had similar chemical compositions (see Table 2). Such behavior indicates that the tailings from different stages of processing of scheelite can be incorporated into masses for the manufacture of red ceramic products without significantly compromising the physical and mechanical properties.

3.5. Carbonation Resistance

Figure 5a–d presents the physical and mechanical properties comparison before and after carbonation of samples sintered at 1000 °C and a heating rate of 5 °C min^{-1}. Such condition was chosen due to the better properties compared to the others. After 3 months of natural aging, the samples significantly reduced physical and mechanical properties (increasing WA and AP and decreasing FS). These changes may be related to the carbonation process. In general, all samples showed similar WA, AP, and FS values, except for M5S2 sample that presented the highest FS value and closed those measured before the carbonation study. Under the percentage viewpoint, the M5S2 sample showed less loss of mechanical strength (~3.6%), while the M5S1 sample showed more significant loss (~57%).

Carbonation is a physical-chemical process in which the CO_2 present in the atmosphere reacts with calcium hydroxide to form calcium carbonate, precipitating inside the microstructure [25]. This process can cause variation in porosity due to the destruction of the mineral through the action of carbonic acid-producing irregular surfaces and an increase in macro and mesoporosity. The precipitation of carbonates fills the micropores and increases the nanoporosity [29,46].

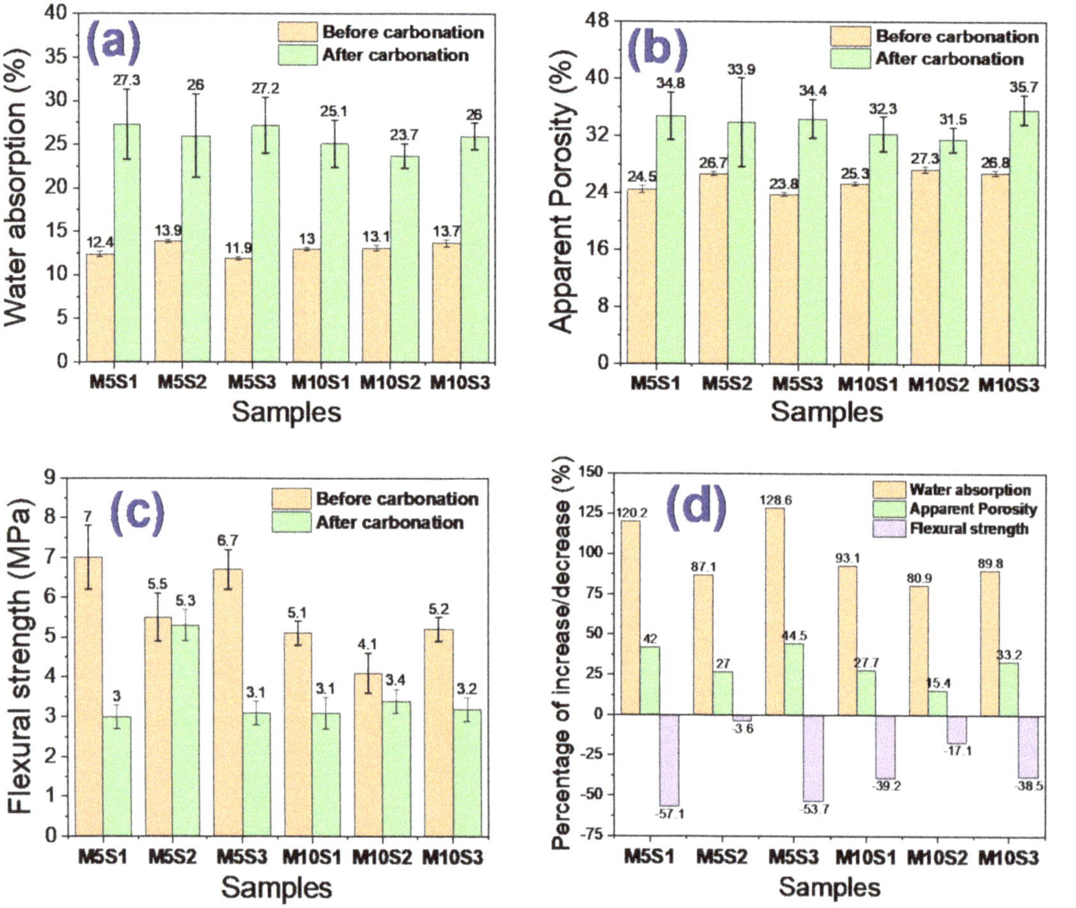

Figure 5. Physical and mechanical properties of the samples fired at 1000 °C (heating rate equal to 5 °C min^{-1}) before and after carbonation: water absorption (**a**), apparent porosity (**b**), flexural strength (**c**), and percentage of increase/decrease (**d**). The carbonation was evaluated after 3 months of natural aging.

In this study, the increase in porosity may be related to the dissolution of the material and the decrease in the flexural strength associated with new asymmetric crystal growth that can generate tensions within the material that contributes to the formation of the cracks. Therefore, pieces obtained with ceramic masses containing tailings rich in CaO (as in the case of scheelite tailings) must be evaluated after the fired step, as the carbonation phenomenon may compromise its function over time.

The X-ray diffraction patterns of samples submitted to the natural aging test for 3 months are shown in Figure 6. The following crystalline phases were observed: quartz (SiO_2, JCPDS 46-1045), anorthite ($CaAl_2Si_2O_8$, JCPDS 89-1472), calcite ($CaCO_3$, JCPDS 72-1937), and monohydrocalcite ($CaCO_3.H_2O$, JCPDS 83-1923). Compared with the diffractograms present in Figure 4c, it is possible to notice an increase in the intensity of the peaks in the region where the anorthite phase is located, probably due to a new phase of monohydrocalcite. The monohydrocalcite from the beginning of the carbonation process will be transformed into carbonate to form calcite later. The newly formed calcite arises from the reaction of the CO_2 present in the environment with the calcium hydroxide contained in the samples. It is probably still present in the samples after the firing step; however, the peaks were not identified due to the low quantity or amorphous form.

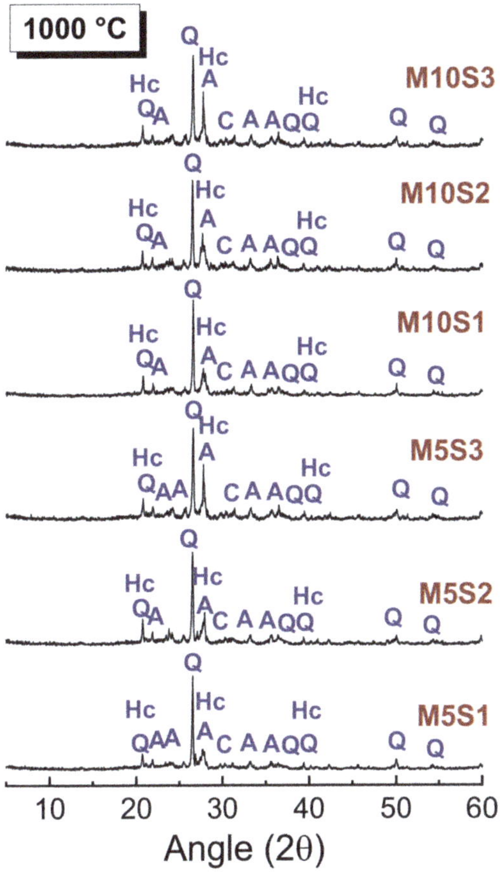

Figure 6. XRD patterns measured from samples sintered at 1000 °C and heating rate of 5 °C min^{-1} after 3 months of natural aging in an internal laboratory environment. Q—Quartz; A—Anorthite; C—calcite; Hc—Monohydrocalcite.

Figure 7 shows the SEM images acquired from the fractured surface of the M5S1, M5S2, and M5S3 samples, sintered at 1000 °C and heating rate of 5 °C min^{-1}, before and after carbonation. It is possible to visualize particle agglomeration of different sizes and shapes, as well as quartz grains. Spherical particles are present in all samples (red circles in the images). Those particles are probably relative to the anorthite phase, confirming the peaks in Figures 4 and 6. After carbonation, the samples presented more pores on their surface. Furthermore, needles were observed (squares highlighted in blue in the images). Those needles are probably relative to the monohydrocalcite phase, which corroborates with the XRD peaks identified in Figure 6. A similar structure was observed by Martín et al. [46].

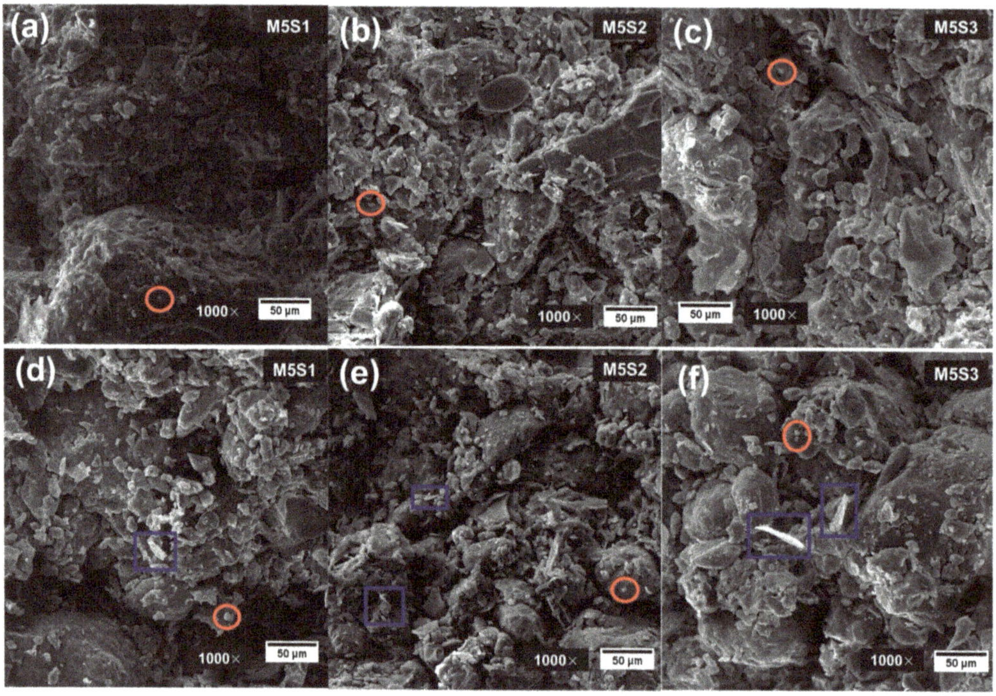

Figure 7. Scanning Electron Microscopy images of samples before (**a**,**c**,**e**) and after (**b**,**d**,**f**) carbonation.

4. Conclusions

Based on the results of this study, it is possible to conclude that incorporating scheelite tailings in ceramic masses for application in red ceramics is a viable alternative. After studying the firing parameters, the following conclusions can be drawn:

- The DRX standards of the 900 and 1000 °C sintered samples identified the gehlenite and anorthite phases, which contributed to the increase of the mechanical resistance of the materials.
- No significant differences were observed in the physical and mechanical properties as a function of the different types of scheelite tailings (ST1, ST2, and ST3) incorporated into the ceramic masses.
- The best physical and mechanical properties (lower water absorption and porosity values and higher flexural strength values) were obtained for samples with 5% scheelite tailings and heated at a rate of 5 °C min^{-1}. For samples with 5% tailings and sintered at 1000 °C, the increase in the heating rate from 5 to 10 °C min^{-1} did not significantly compromise the properties.
- Samples exposed to ambient conditions for 3 months showed a loss of physical and mechanical properties, probably due to the onset of the carbonation phenomenon. The M5S2 and M10S2 samples were the ones that presented the lowest percentages of resistance loss.
- The incorporation of scheelite tailings presented the potential for application in red ceramics and an alternative to reduce environmental pollution and conserve mineral resources.

Author Contributions: M.E.A.C., V.J.d.S. and E.P.d.A.B.: Methodology, formal analysis, and investigation, writing—original draft preparation; A.M.R. and F.P.d.C.: Formal analysis and investigation, writing—original draft preparation, writing—review and editing; R.R.M., G.A.N. and L.N.d.L.S.: Conceptualization, formal analysis, funding acquisition, resources, supervision. All authors have read and agreed to the published version of the manuscript.

Funding: This research was funded by Coordenação de Aperfeiçoamento de Pessoal de Nível Superior (CAPES), grant number 88887.597478/2021-00, and by Conselho Nacional de Desenvolvimento Científico e Tecnológico (CNPq).

Institutional Review Board Statement: Not applicable.

Informed Consent Statement: Not applicable.

Data Availability Statement: Not applicable.

Acknowledgments: The authors would like to thank the support provided by the Laboratório de Tecnologia dos Materiais, where it was possible to carry out the experiments of this work.

Conflicts of Interest: The authors declare no conflict of interest.

References

1. Almeida, E.P.; Carreiro, M.E.A.; Rodrigues, A.M.; Ferreira, H.S.; Santana, L.N.L.; Menezes, R.R.; Neves, G.A. A new eco-friendly mass formulation based on industrial mining residues for the manufacture of ceramic tiles. *Ceram. Int.* **2021**, *47*, 11340–11348. [CrossRef]
2. Da Costa, F.P.; da Morais, C.R.S.; Rodrigues, A.M. Sustainable glass-ceramic foams manufactured from waste glass bottles and bentonite. *Ceram. Int.* **2020**, *46*, 17957–17961. [CrossRef]
3. Da Silva, A.L.; Luna, C.B.B.; de Farias, A.F.F.; de Medeiros, S.A.S.L.; Meneghetti, S.M.P.; Rodrigues, A.M.; de Costa, A.C.F.M. From Disposal to Reuse: Production of Sustainable Fatty Acid Alkyl Esters Derived from Residual Oil Using a Biphasic Magnetic Catalyst. *Sustainability* **2020**, *12*, 10159. [CrossRef]
4. Da Silva, A.L.; Farias, A.F.F.; Pontes, J.R.M.; Rodrigues, A.M.; de Costa, A.C.F.M. Synthesis of the ZnO-Ni0.5Zn0.5Fe2O4-Fe2O3 magnetic catalyst in pilot-scale by combustion reaction and its application on the biodiesel production process from oil residual. *Arab. J. Chem.* **2020**, *13*, 7665–7679. [CrossRef]
5. Fernandes, J.V.; Guedes, D.G.; da Costa, F.P.; Rodrigues, A.M.; de Neves, G.A.; Menezes, R.R.; de Santana, L.N.L. Sustainable Ceramic Materials Manufactured from Ceramic Formulations Containing Quartzite and Scheelite Tailings. *Sustainability* **2020**, *12*, 9417. [CrossRef]
6. De Figueirêdo, J.M.R.; da Costa, F.P.; Fernandes, J.V.; Rodrigues, A.M.; de Neves, G.A.; Menezes, R.R.; de Santana, L.N.L. Development of Scheelite Tailings-Based Ceramic Formulations with the Potential to Manufacture Porcelain Tiles, Semi-Stoneware and Stoneware. *Materials* **2020**, *13*, 5122. [CrossRef] [PubMed]
7. Rehman, M.; Ahmad, M.; Rashid, K. Influence of fluxing oxides from waste on the production and physico-mechanical properties of fired clay brick: A review. *J. Build. Eng.* **2020**, *27*, 100965. [CrossRef]
8. Mymrin, V.; Alekseev, K.; Catai, R.; Nagalli, A.; Aibuldinov, Y.; Bekturganov, N.S.; Izzo, R.L. Red ceramics from composites of hazardous sludge with foundry sand, glass waste and acid neutralization salts. *J. Environ. Chem. Eng.* **2016**, *4*, 753–761. [CrossRef]
9. Xu, Z.; Li, S.; Meng, Q.; Wang, X.; Zhu, Q.; Li, J.G. Generalized synthesis of NaLn(MoO4)2 nano/microcrystals (Ln = La–Lu and Y): The effects of lanthanide contraction, structure, and down-/up-conversion luminescence. *J. Alloys Compd.* **2020**, *830*, 154676. [CrossRef]
10. De Brito, I.P.; de Almeida, E.P.; de Araújo Neves, G.; de Lucena Lira, H.; Menezes, R.R.; da Silva, V.J.; de Lima Santana, L.N. Development of cordierite/mullite composites using industrial wastes. *Int. J. Appl. Ceram. Technol.* **2021**, *18*, 253–261. [CrossRef]
11. Loutou, M.; Misrar, W.; Koudad, M.; Mansori, M.; Grase, L.; Favotto, C.; Taha, Y.; Hakkou, R. Phosphate mine tailing recycling in membrane filter manufacturing: Microstructure and filtration suitability. *Minerals* **2019**, *9*, 318. [CrossRef]
12. Fontes, W.C.; de Carvalho, J.M.F.; Andrade, L.C.R.; Segadães, A.M.; Peixoto, R.A.F. Assessment of the use potential of iron ore tailings in the manufacture of ceramic tiles: From tailings-dams to "brown porcelain". *Constr. Build. Mater.* **2019**, *206*, 111–121. [CrossRef]
13. Liu, S.; Guan, X.; Zhang, S.; Dou, Z.; Feng, C.; Zhang, H.; Luo, S. Sintered bayer red mud based ceramic bricks: Microstructure evolution and alkalis immobilization mechanism. *Ceram. Int.* **2017**, *43*, 13004–13008. [CrossRef]
14. Xu, X.; Song, J.; Li, Y.; Wu, J.; Liu, X.; Zhang, C. The microstructure and properties of ceramic tiles from solid wastes of Bayer red muds. *Constr. Build. Mater.* **2019**, *212*, 266–274. [CrossRef]
15. Drif, B.; Taha, Y.; Hakkou, R.; Benzaazoua, M. Integrated valorization of silver mine tailings through silver recovery and ceramic materials production. *Miner. Eng.* **2021**, *170*, 107060. [CrossRef]
16. Karhu, M.; Lagerbom, J.; Solismaa, S.; Honkanen, M.; Ismailov, A.; Räisänen, M.L.; Huttunen-Saarivirta, E.; Levänen, E.; Kivikytö-Reponen, P. Mining tailings as raw materials for reaction-sintered aluminosilicate ceramics: Effect of mineralogical composition on microstructure and properties. *Ceram. Int.* **2019**, *45*, 4840–4848. [CrossRef]

17. Silva, M.C.A.; Leão, V.A.; Reis, E.L. Incorporation of quartzite fines in the production of red ceramics. *J. Clean. Prod.* **2021**, *288*, 125098. [CrossRef]
18. Feng, B.; Guo, W.; Peng, J.; Zhang, W. Separation of scheelite and calcite using calcium lignosulphonate as depressant. *Sep. Purif. Technol.* **2018**, *199*, 346–350. [CrossRef]
19. Chen, W.; Feng, Q.; Zhang, G.; Yang, Q. Investigations on flotation separation of scheelite from calcite by using a novel depressant: Sodium phytate. *Miner. Eng.* **2018**, *126*, 116–122. [CrossRef]
20. Yeşilay, S.; Çakı, M.; Ergun, H. Usage of marble wastes in traditional artistic stoneware clay body. *Ceram. Int.* **2017**, *43*, 8912–8921. [CrossRef]
21. Traore, K.; Kabre, T.; Blanchart, P. Gehlenite and anorthite crystallisation from kaolinite and calcite mix. *Ceram. Int.* **2003**, *29*, 377–383. [CrossRef]
22. Reis, G.; Cazacliu, B.; Artoni, R.; Torrenti, J.M. Effect of the accelerated carbonation treatment on the recycled sand physicochemical characteristics through the rolling carbonation process. *J. CO2 Util.* **2020**, *39*, 101181. [CrossRef]
23. De Santana, L.L.; da Silva, B.J.; Gonçalves, W.P.; Cartaxo, J.M.; Lira, B.S.; dos Santos, R.C.; Menezes, R.R. Influence of firing conditions on properties of red ceramic. *Mater. Sci. Forum* **2012**, *727*, 721–726. [CrossRef]
24. Mahmoudi, S.; Bennour, A.; Srasra, E.; Zargouni, F. Characterization, firing behavior and ceramic application of clays from the Gabes region in South Tunisia. *Appl. Clay Sci.* **2017**, *135*, 215–225. [CrossRef]
25. Kong, L.; Han, M.; Yang, X. Evaluation on relationship between accelerated carbonation and deterioration of concrete subjected to a high-concentrated sewage environment. *Constr. Build. Mater.* **2020**, *237*, 117650. [CrossRef]
26. Basu, P.; Gupta, R.C.; Agrawal, V. Effect of Carbonation on the Mechanical and Durability Properties of Sandstone Modified Self-Compacting Concrete. In *Materials Today Proceedings*; Elsevier Ltd.: Amsterdam, The Netherlands, 2020; Volume 32, pp. 686–691.
27. Li, T.; Wang, S.; Xu, F.; Meng, X.; Li, B.; Zhan, M. Study of the basic mechanical properties and degradation mechanism of recycled concrete with tailings before and after carbonation. *J. Clean. Prod.* **2020**, *259*, 120923. [CrossRef]
28. Mi, R.; Pan, G.; Liew, K.M. Predicting carbonation service life of reinforced concrete beams reflecting distribution of carbonation zones. *Constr. Build. Mater.* **2020**, *255*, 119367. [CrossRef]
29. Martín, D.; Aparicio, P.; Galán, E. Time evolution of the mineral carbonation of ceramic bricks in a simulatedpilot plant using a common clay as sealing material at superficial conditions. *Appl. Clay Sci.* **2019**, *180*, 1–11. [CrossRef]
30. Machado, T.G.; Gomes Uílame, U.; Monteiro, F.M.; Valcacer, S.M.; da Silva Gilson, G. Analysis of the incorporation of scheelite residue in kaolinitic clay of Boa Saúde -RN. *Mater. Sci. Forum* **2012**, *727*, 844–849. [CrossRef]
31. Casagrande, A. Notes on the design of the liquid limit device. *Geotechnique* **1958**, *8*, 84–91. [CrossRef]
32. Da Costa, F.P.; da Morais, C.R.S.; Pinto, H.C.; Rodrigues, A.M. Microstructure and physico-mechanical properties of Al2O3-doped sustainable glass-ceramic foams. *Mater. Chem. Phys.* **2020**, *256*, 123612. [CrossRef]
33. Song, J.L.; Lee, G.Y.; Choi, J.P.; Lee, J.S. Compaction behavior of bimodal iron nanopowder agglomerate. *Powder Technol.* **2018**, *338*, 333–341. [CrossRef]
34. Menezes, R.; Ferreira, H.; Neves, G.; Lira, H.D.L.; Ferreira, H.C. Use of granite sawing wastes in the production of ceramic bricks and tiles. *J. Eur. Ceram. Soc.* **2005**, *25*, 1149–1158. [CrossRef]
35. McConville, C.J.; Lee, W.E. Microstructural development on firing illite and smectite clays compared with that in kaolinite. *J. Am. Ceram. Soc.* **2005**, *88*, 2267–2276. [CrossRef]
36. Torres, P.; Manjate, R.S.; Quaresma, S.; Fernandes, H.R.; Ferreira, J.M.F. Development of ceramic floor tile compositions based on quartzite and granite sludges. *J. Eur. Ceram. Soc.* **2007**, *27*, 4649–4655. [CrossRef]
37. Baccour, H.; Medhioub, M.; Jamoussi, F.; Mhiri, T. Influence of firing temperature on the ceramic properties of Triassic clays from Tunisia. *J. Mater. Process. Technol.* **2009**, *209*, 2812–2817. [CrossRef]
38. Jiang, F.; Li, Y.; Zhao, L.; Cang, D. Novel ceramics prepared from inferior clay rich in CaO and Fe2O3: Properties, crystalline phases evolution and densification process. *Appl. Clay Sci.* **2017**, *143*, 199–204. [CrossRef]
39. Gonzáles-García, F.; Romero-Acosta, V.; García-Ramos, G.; Gonzáles-Rodríguez, M. Firing transformations of mixtures of clays containing illite, kaolinite and calcium carbonate used by ornamental tile industries. *Appl. Clay Sci.* **1990**, *5*, 361–375. [CrossRef]
40. Ouahabi, M.; Daoudi, L.; Hatert, F.; Fagel, N. Modified mineral phases during clay ceramic firing. *Clays Clay Miner.* **2015**, *65*, 404–413. [CrossRef]
41. Ptáček, P.; Opravil, T.; Šoukal, F.; Havlica, J.; Holešinsk, R. Kinetics and mechanism of formation of gehlenite, Al–Si spinel and anorthite from the mixture of kaolinite and calcite. *Solid State Sci.* **2013**, *26*, 53–58. [CrossRef]
42. ASTM. *C20-00, Standard Test Methods for Apparent Porosity, Water Absorption, Apparent Specific Gravity, and Bulk Density of Burned Refractory Brick and Shapes by Boiling Water*; ASTM Int.: West Conshohocken, PA, USA, 2015. [CrossRef]
43. ASTM. *C67/C67M-20, Standard Test Methods for Sampling and Testing Brick and Structural Clay Tile*; ASTM Int.: West Conshohocken, PA, USA, 2020. [CrossRef]
44. De Faria, J.; Manhães, R.; da Luz, F.S.; Monteiro, S.N.; Vieira, C.M.F. Incorporation of unserviceable tire waste in red ceramic. *J. Mater. Res. Technol.* **2019**, *8*, 6041–6050. [CrossRef]

45. Coletti, C.; Maritan, L.; Cultrone, G.; Hein, A.; Molina, E.; Mazzoli, C. Recycling trachyte waste from the quarry to the brick industry: Effects on physical and mechanical properties, and durability of new bricks. *Constr. Build. Mater.* **2018**, *166*, 792–807. [CrossRef]
46. Martín, D.; Aparicio, P.; Galan, E. Accelerated carbonation of ceramic materials. Application to bricks from Andalusian factories (Spain). *Constr. Build. Mater.* **2018**, *181*, 598–608. [CrossRef]

sustainability

MDPI

Article

Development of Eco-Friendly Mortars Produced with Kaolin Processing Waste: Durability Behavior Viewpoint

Alisson Mendes Rodrigues [1,*], Fabiana Pereira da Costa [2], Suellen Lisboa Dias Beltrão [2], Jucielle Veras Fernandes [2], Romualdo Rodrigues Menezes [1] and Gelmires de Araújo Neves [1]

[1] Laboratory of Materials Technology (LTM), Department of Materials Engineering, Federal University of Campina Grande, Bodocongó, Campina Grande 58429-900, Brazil; romualdo.menezes@ufcg.edu.br (R.R.M.); gelmires.neves@ufcg.edu.br (G.d.A.N.)

[2] Graduate Program in Materials Science and Engineering (PPG-CEMat), Department of Materials Engineering, Federal University of Campina Grande, Bodocongó, Campina Grande 58429-900, Brazil; fabiana.costa@estudante.ufcg.edu.br (F.P.d.C.); slisboabeltrao@gmail.com (S.L.D.B.); jucielle_fernandes@hotmail.com (J.V.F.)

* Correspondence: alisson.mendes@professor.ufcg.edu.br

Abstract: This study presents the development of new eco-friendly mortar compositions containing kaolin residues (KR) and assesses their durability behavior. Firstly, the natural and calcinated kaolin residues (600 °C, 650 °C, 700 °C, 750 °C, and 800 °C) were characterized by X-ray diffraction (XRD), differential thermal analysis (DTA), granulometric analysis, and surface area. The kaolin residue calcinated at 800 °C was chosen to be added to new compositions of mortar because it presented the best pozzolanic performance. The aging tests accomplished in internal (E_i) and external (E_e) environments were applied in mortars with a mass proportion of 1:2:6 (cement + KR: lime: sand), in which the KR, calcinated at 800 °C, replaced the cement in the mass fraction of 0%, 5%, 10%, 15%, 20%, and 30%. The E_i was performed for 30, 60, 90, 180, and 360 days, and the E_e for 90; 210; 360; and 512 days. After the aging tests were completed, the mortar compositions containing KR were evaluated to determine their mineralogical phases (XRD), compressive strength (CS), and thermal behavior (DTA and thermogravimetry). In summary, the KR addition to the mortar compositions decreases the mechanical resistance to compression; however, mortars with a substitution of 10% and 20% presented resistance values within the minimum limit of 2.4 MPa established by ASTM C 270.

Keywords: kaolin; solid residues; pozzolanic activity; cement mortar; carbonation

Citation: Rodrigues, A.M.; Costa, F.P.d.; Beltrão, S.L.D.; Fernandes, J.V.; Menezes, R.R.; Neves, G.d.A. Development of Eco-Friendly Mortars Produced with Kaolin Processing Waste: Durability Behavior Viewpoint. *Sustainability* **2021**, *13*, 11395. https://doi.org/10.3390/su132011395

Academic Editor: Jorge de Brito

Received: 9 August 2021
Accepted: 9 October 2021
Published: 15 October 2021

1. Introduction

Sustainable construction materials are being increasingly studied to minimize the environmental impacts generated by the civil construction sector. Many studies have targeted alternative materials to replace the use of cement in concretes and mortars to improve their sustainability. This is because the Portland cement production process releases one of the highest CO_2 emission rates in the world. It is estimated that about one ton of carbon dioxide is created for each ton of Portland cement produced [1–3].

Several studies have been carried out using alternative cementitious materials to replace Portland cement in concretes and mortars to mitigate their environmental impacts and reduce costs. Among the alternative materials are fly ash [4–6], blast furnace slag [7–9], eggshell powder [2,10], silica fume [11–13], limestone [14,15], calcium carbide residue [16], and calcined kaolin residue, which is undoubtedly one of the most well known alternative cementitious materials [17]. The kaolinite clay is the main constituent in the kaolin residues. When such residues are submitted to thermal treatment, under controlled conditions, kaolinite dehydroxylation is guaranteed, whereby it transforms itself into an amorphous structure called metakaolinite. Metakaolin can be used in civil construction and added to mortars to act as a material pozzolanic. The main property of a pozzolana is its ability to react and combine with calcium hydroxide, forming stable compounds with an agglomerating power, such as hydrated calcium silicates and aluminates [17–19].

The positive effects exerted by pozzolans on the mechanical properties and durability of mortars and concretes have been emphasized in several studies. Al-Akhras [20] investigated the sulfate resistance of concretes produced from the mixture of metakaolinite (MK) with Portland cement (PC), at the substitution levels of 5, 10, and 15 wt%, and concluded that such a substitution was beneficial for the resistance to the sulfate, which increases with the increase in MK content in the mixture. Guneyisi et al. [21] used MK to substitute Portland cement at levels of 10 and 20 wt% to improve the performance of the concrete. The authors evaluated the durability properties of samples produced from water absorption (WA), drying shrinkage (DS), and porosity (P) measures, and reported that the incorporation of MK reduced the drying shrinkage deformation and improved the pore structures of the samples. In addition, they observed a decreasing trend in WA with an increase in the MK content.

It is worth mentioning that durability evaluation is undoubtedly the most critical aspect of cementitious material development because, for it to be reliable, knowledge of its conditions of use and its useful life is essential [22–24]. Although the calcined kaolin residue has been extensively studied as a pozzolanic material, studies that addressed the long-term durability behavior of mortars containing kaolin residue are still scarce. Therefore, this work aims to use calcined kaolin residue as a partial substitute for cement in different percentages (0%, 5%, 10%, 15%, 20%, and 30%) in the composition of mortar for coating, and to assess the overall impact on the long-term durability behavior. Mortar durability will be evaluated by natural aging tests in the internal (30, 60, 90, 180, and 360 days) and external (90, 210, 360, and 512 days) environments.

2. Materials and Methods

2.1. Raw Materials

The Portland cement type CPII-F 32 (PC) was used as a binder following the ASTM C150 standard (CIMPOR BRASIL) [25]. This cement was used because it does not have the addition of pozzolan. The other materials used were hydrated lime (HL) (Carbomil); sand (S), with a density of 2.6 g/cm^3 and an average particle diameter of 4.8 mm; and kaolin residue (KR) supplied by Caulim Caiçara S/A (Brazil). The as-received kaolin residue was pulverized in a MultiNo® (270 M/S/M) compressed air mill for 30 min, and heat-treated for 2 h at different temperatures (600 °C, 650 °C, 700 °C, 750 °C, and 800 °C). The heat treatments were carried out in an electric oven (Flyever Equipment, FE 50 RP) at a heating rate of 5 °C/min^{-1}.

2.2. Preparation and Curing of the Mortar Samples

Samples with a mass proportion of 1:2:6 (cement + KR: lime: sand) were prepared with and without the kaolin residue. In the samples with kaolin residue, the cement was replaced. The mass fractions of Portland cement substitutions by calcined kaolin residue were 0%, 5%, 10%, 15%, 20%, and 30%.

The raw materials were mixed in a mechanical planetary motion mixer for 5 min. After that, water was added to the mixture until the consistency index was 260 ± 10mm (ASTM C 1437) [26]. The test specimens were molded in a cylindrical mold with the dimensions of 50 mm × 100 mm (diameter × height). After 24 h, the samples were demolded and immersed in water with lime (2% concentration) for 30 days. This procedure was carried out to avoid carbonation. Table 1 provides details of the proportions used for the mortars and the nominal composition of replaced cement.

2.3. Mortar Durability Assessment

Aging tests in internal (E_i) (laboratory) and external (E_e) environments were carried out to evaluate the durability of the two mortar compositions studied. All experiments were carried out in triplicate. The aging test specimens in an internal environment were kept in a place protected from external climatic actions (a laboratory environment with temperatures between 21 °C and 30 °C, and an air humidity of ~78%) for 30, 60, 90, 180, and 360 days. The test specimens submitted to the aging test in an external environment were kept in a place that permitted their exposure to the environment's natural climatic conditions. The samples from this test were kept in this environment for 90, 210, 360, and 512 days.

Table 1. Nominal composition and percentages of cement replaced by calcined kaolin residue.

Designation	Percentage of PC Replaced by KR (%)	
	PC	KR
M0	100	-
M5	95	5
M10	90	10
M15	85	15
M20	80	20
M30	70	30

2.4. *Characterizations*

The chemical composition was defined using X-ray fluorescence (Shimadzu, EDX 720). The mineralogical phase characterization was performed by X-ray diffraction (XRD) using a Shimadzu XRD6000 equipment with Cu Kα radiation (40kV/30mA), and a goniometer speed of 2 °C/min^{-1}, step of 0.02°, and the JCPDS database. Thermogravimetric (TGA) and differential thermal (DTA) analyses were performed using a BP Engenharia equipment (model RB-300) with a nitrogen atmosphere and heating rate equal to 12.5 °C/min^{-1}. All thermal analyses were performed in platinum crucibles and the temperature range was kept between 25–1000 °C. Before the mineralogical (XRD) and thermal (TGA and DTA) tests, the mortar samples were packed in polyethylene bags to prevent carbonation and the tests were carried out on the same day.

The particle size distribution of kaolin residue was determined by laser diffraction (Cilas, 1064 LD), and the specific area was evaluated using the BET methodology (ASAP 2420 V2.02). The pozzolanic activity (I_{PA}) was measured from test specimens (50 mm x 100 mm) manufactured from two mortar compositions. The first composition, called reference mortar (mortar ref.), only contained Portland cement. In the second composition (mortar KR), 20% of the cement was replaced by calcined kaolin residue. In both compositions, the H_2O/PC or H_2O/(PC + KR) ratio was equal to 0.485, as specified in ASTM C311M [27]. Compressive strength (CS) measurements were performed using test specimens cured for 7, 14, and 28 days, and the I_{PA} value was calculated with the aid of Equation (1):

$$I_{PA}(\%) = \frac{CSM_{KR}}{CSM_{Ref}} \times 100 \quad (1)$$

where CSM_{Ref} (MPa) and CSM_{KR} (MPa) correspond to the CS values measured using the mortar ref. and mortar KR samples, respectively.

After each exposure period (durability tests), compressive strength tests were carried out in accordance with ASTM C39/C39M [28]. All experiments were performed on a universal mechanical testing machine (SHIMADZU, AG-IS) with a loading speed of 0.25 ± 0.05 MPa/s. Mineralogical (XRD) and thermal (TGA and DTA) tests were also carried out to monitor the phase transformations.

3. Results and Discussions

3.1. *Raw Materials Characterizations*

The chemical compositions of Portland cement type CPII-F 32, hydrated lime, and kaolin residue are listed in Table 2. Figure 1a,b show XRD patterns of Portland cement type CPII-F 32 and of hydrated lime. The mineralogical phases: tricalcium silicate (C3S—$3CaO{\cdot}SiO_2$,

JCPDS 49-0442); dicalcium silicate (C2S—$2CaO{\cdot}SiO_2$, JCPDS 33-0302); tricalcium aluminate (C3A—$3CaO{\cdot}Al_2O_3$, JCPDS 38-1429); iron tetracalcium aluminate (C4AF—$4CaO{\cdot}Al_2O_3{\cdot}Fe_2O_3$, JCPDS 30-0226); calcium carbonate ($CaCO_3$, JCPDS 05-0586); and calcium sulfate hemihydrate ($CaSO_4{\cdot}1/2H_2O$, JCPDS 0041-0244), were detected in the Portland cement (Figure 1a). In the hydrated lime (Figure 1b) the phases of portlandite ($Ca(OH)_2$, JCPDS 72-0156) and calcite ($CaCO_3$, JCPDS 05-0586) were detected.

Table 2. The chemical composition of Portland cement type CPII-F 32, hydrated lime, and kaolin residue.

Raw Materials	Oxides								
	SiO_2	Al_2O_3	Fe_2O_3	K_2O	MgO	CaO	SO_3	Others	LOI [1]
PC	19.6	10.3	1.1	-	11.5	38.0	8.3	1.5	9.7
HL	1.6	0.6	0.1	0.3	1.6	72.3	-	0.4	23.1
KR	58.8	36.0	0.2	1.6	2.2	-	-	0.4	0.8

[1] LOI—Loss on ignition measured after drying at 110 °C and firing at 1000 °C.

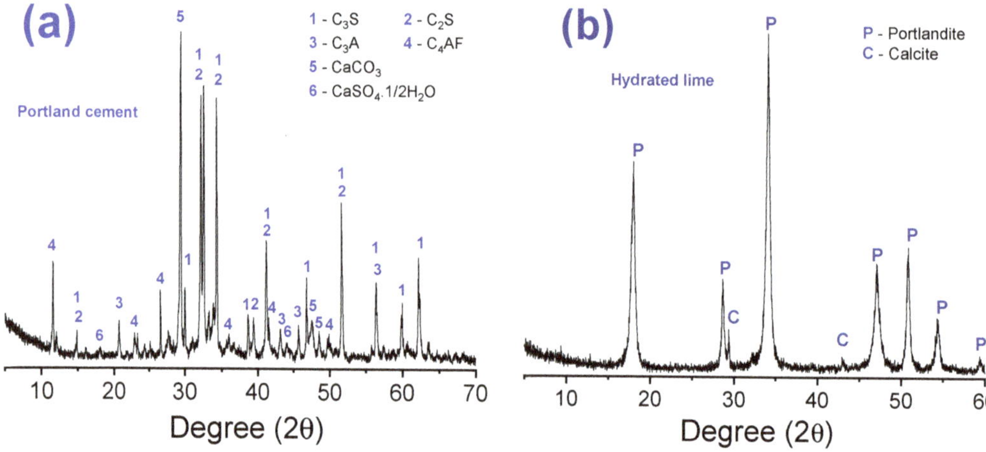

Figure 1. The XRD patterns measured from Portland cement type CPII-F 32 (**a**) and hydrated lime (**b**).

Figure 2a–f show the XRD patterns of natural and calcined kaolin residues at different temperatures (600 °C, 650 °C, 700 °C, 750 °C, and 800 °C). The mineralogical phases: quartz (SiO_2, JCPDS 46-1045); mica ($KMg_3(Si_3Al)O_{10}(OH)_2$, JCPDS 83-1808); and kaolinite ($Al_2Si_2O_5(OH)_4$, JCPDS 14-0164) were detected in the natural kaolin residue. Except for kaolinite, all other mineralogical phases were also identified in the calcined kaolin residues. At temperatures above 500 °C, the kaolinite suffers dehydroxylation with the consequent breakdown of the structure, which then forms the metakaolinite ($Al_2Si_2O_5(OH)_4 \rightarrow Al_2O_3{\cdot}2SiO_2 + 2H_2O$). However, the characteristic amorphous band of metakaolin is not evident in Figure 2b–f, due to the relatively high intensity of the quartz peak ($2\theta = 26.64°$).

The DTA curves measured from natural and calcined kaolin residues are shown in Figure 3. The natural kaolin residue has an endothermic peak at around 600 °C, which is characteristic of kaolinite dehydroxylation, giving rise to amorphous metakaolinite [29]. Such a kaolinite dehydroxylation peak (at 600 °C) was not observed in the DTA curves measured from the calcined kaolinite residues. An exothermic peak at 900 °C is related to mullite nucleation [30,31], which was identified in all DTA curves.

The granulometric and surface area analysis of Portland cement particles, hydrated lime, and natural and calcined kaolin residues are listed in Table 3. In general, the natural and calcined kaolin residues presented an average particle diameter between 12.1 µm and 19.0 µm. The specific surface area is a relevant factor in the pozzolanic reaction since it depends strongly on the structure of the surface and its specific area [32,33]. It is worth remembering that the larger the specific area, the greater the reactivity of the material. Furthermore, the material's specific area tends to be larger if the particle size is smaller [34,35]. A greater fineness of the kaolin residue leads to an improvement in its pozzolanicity. Therefore, the calcined kaolin residue at the temperature of 800 °C was used to prepare the mortar compositions investigated in this study because it had the smallest average particle diameter (12.8 µm) and the largest specific area (25.95 m^2/g). Huang et al. [29] evaluated the influence of the calcination temperature (500–1000 °C) on the pozzolanic activity in the kaolin residues from southwest China. They found that the best calcination temperature for converting the kaolin residue into metakaolin with a high pozzolanic index was 800 °C.

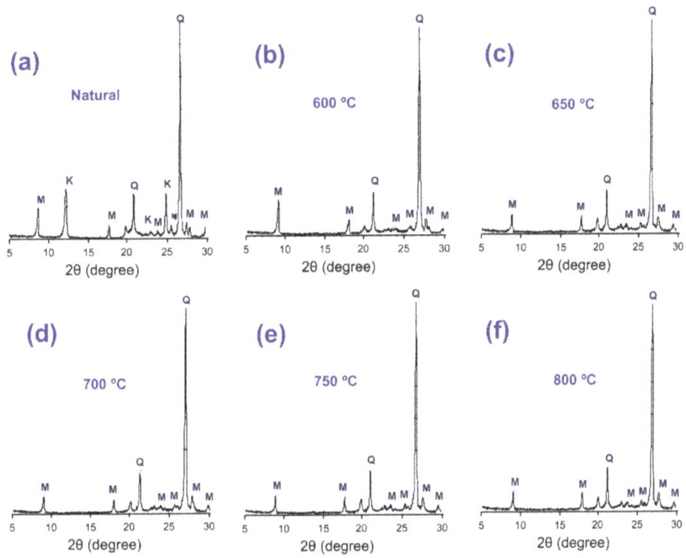

Figure 2. The XRD patterns measured from natural kaolin (**a**) and calcined residues at 600 °C (**b**), 650 °C (**c**), 700 °C (**d**), 750 °C (**e**), and 800 °C (**f**). (M: Mica, Q: Quartz, and K: Kaolinite).

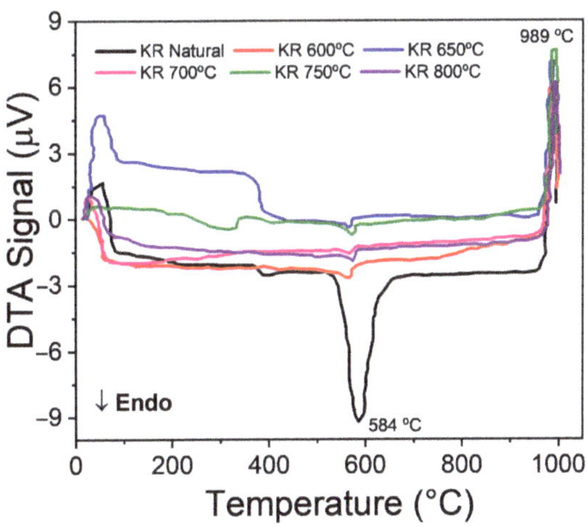

Figure 3. The DTA curves of natural and calcined kaolin residues at 600 °C, 650 °C, 700 °C, 750 °C, and 800 °C.

Table 3. The particle diameters of the Portland cement (PC), hydrated lime (HL), and natural and calcined kaolin residues (KR) at 600 °C, 650 °C, 700 °C, 750 °C, and 800 °C.

Diameter and Specific Area	Raw Materials							
	PC	HL	KR Natural	KR 600 °C	KR 650 °C	KR 700 °C	KR 750 °C	KR 800 °C
D_{10} (μm)	1.8	0.8	1.4	2.9	1.6	2.8	2.6	1.5
D_{50} (μm)	13.8	5.7	9.3	14.3	10.1	14.1	13.8	9.4
D_{90} (μm)	43.9	21.1	25.8	42.7	29.7	42.5	42.8	27.8
D_m (μm)	19.1	9.0	12.1	19.0	13.6	18.8	18.7	12.8
Specific area (m^2/g)	-	-	29.73	24.98	23.18	25.10	20.52	25.95

D_{10} = Diameter at 10%; D_{50} = Diameter at 50%; D_{90} = Diameter at 90%; and D_m = Average diameter.

Figure 4 shows the compressive strength (CS) and the pozzolanic activity index (I_{PA}) measured from mortars without and with kaolin residue calcined at 800 °C (mortar ref. and mortar KR 800 °C, respectively). As previously mentioned, in the sample used to assess the I_{PA} value, 20% of the cement was replaced by calcined kaolin residue. All samples were cured for 7, 14, and 28 days. For samples cured for 7 days, it was observed that the addition of kaolin residue resulted in a decrease in compressive strength of 8.2% (I_{PA} = 91.8%). Such behavior occurred because replacing part of the cement with pozzolan decreases its resistance, due to less cementing fraction. In addition, pozzolanic reactions are slow, so it is only with time that resistance increases. Such behavior is related to the formation of calcium silicate hydrate (CSH) resulting from the reaction between calcined kaolin residue and portlandite, i.e., a pozzolanic reaction. As expected, the CSH formation increases the mechanical strength. For cure times equal to 14 and 28 days, the mortar KR 800 °C sample showed slight CS values than the samples measured for the mortar ref. (I_{PA} > 100%). From the perspective of mechanical strength, this result indicates that the kaolin residue, calcined at 800 °C, has the potential to replace cement in mortar. All I_{PA} values were higher than the minimum limit of 75% recommended by ASTM C618 [36].

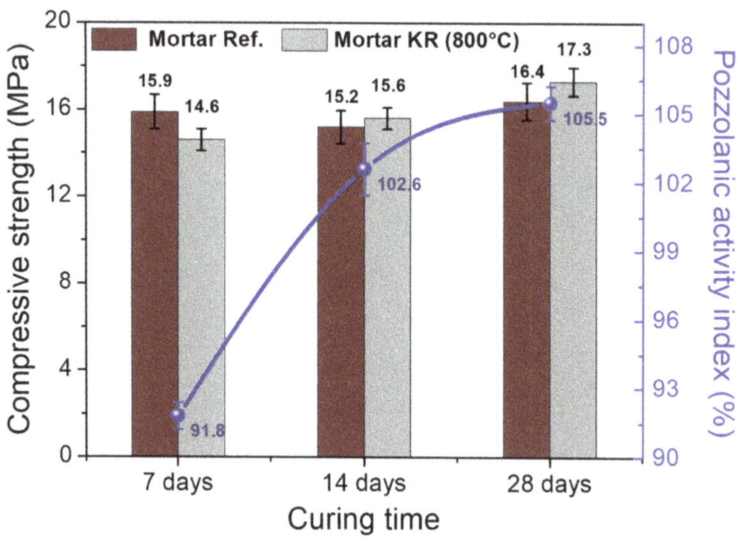

Figure 4. The compressive strength and pozzolanic activity index values for mortar ref. and mortar KR 800 °C samples.

3.2. Mortar Properties after Aging in Internal and External Environments

Figure 5a–e shows XRD patterns of mortar KR 800 °C aged under internal environments (E_i) for 30, 180, and 360 days (Figure 5a–c) and external environments (E_e) for

360 and 512 days (Figure 5d–e). The portlandite ($Ca(OH)_2$, JCPDS 72-0156); ettringite ($3CaO \cdot Al_2O_3 \cdot 3CaSO_4 \cdot 32H_2O$, JCPDS 72-0646); calcite ($CaCO_3$, JCPDS 05-0586); and quartz (SiO_2, JCPDS 46-1045) were phases identified in the mortars exposed to the internal and external environment. However, the intensity of the portlandite phase peaks decreased with time. Such behavior is strongly related to the pozzolanic reactions and the carbonation process. This indicates that the pozzolanic reaction and the carbonation process, over time, have consumed the portlandite.

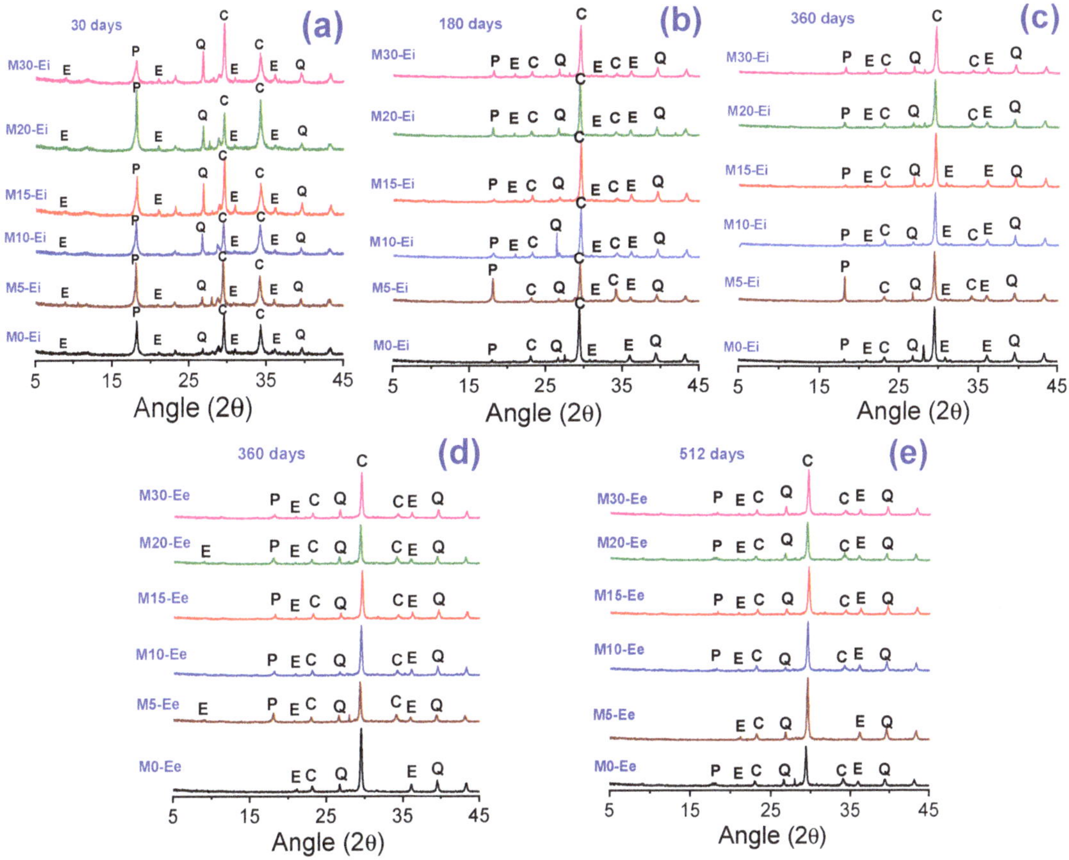

Figure 5. The XRD patterns of aged mortars in the internal environment of the laboratory for 30 days (**a**); 180 days (**b**); and 360 days (**c**), and of the exposure to the external environment for 360 days (**d**) and 512 days (**e**). (P: Portlandite, E: Ettringite, Q: Quartz, and C: Calcite).

Figure 6a–d show the TGA and DTA curves of mortars aged in an internal environment for 30 days (Figure 6a,b) and 360 days (Figure 6c,d). In general, it is possible to see that all the curves have a similar thermic profile. In summary, all mortars show four stages of mass loss. The first stage is identified in the temperature range between 25–100 °C and is associated with the evaporation of free water, adsorbed water loss, and water coordinated to cations. The second stage occurs in the temperature range between 101–400 °C and corresponds to the dehydration of ettringite, hydrated calcium silicate (CSH), and the decomposition of hemi compounds and monocarbonates (AFm). The dehydration regions of CSH and ettringite are superimposed; however, it is possible to observe peaks at around 156 °C which correspond to the loss of water in the ettringite phase [37–39].

The decomposition of hemi and monocarbonate components is more evident at 360 days (Figure 6d), a process in which it is possible to observe the formation of an endothermic band at around 215 °C. The last 2 stages occur in the temperature range between 401–500 °C and 501–1000 °C, and are related to the decomposition of calcium hydroxide ($Ca(OH)_2 \rightarrow CaO + H_2O$) and calcium carbonate ($CaCO_3 \rightarrow CaO + CO_2$), respectively [40–42]. The mass loss in these regions can be used to determine the $Ca(OH)_2$ and $CaCO_3$ content. It is worth mentioning that calcite ($CaCO_3$), a carbonation product, tends to reduce the mortars' strength.

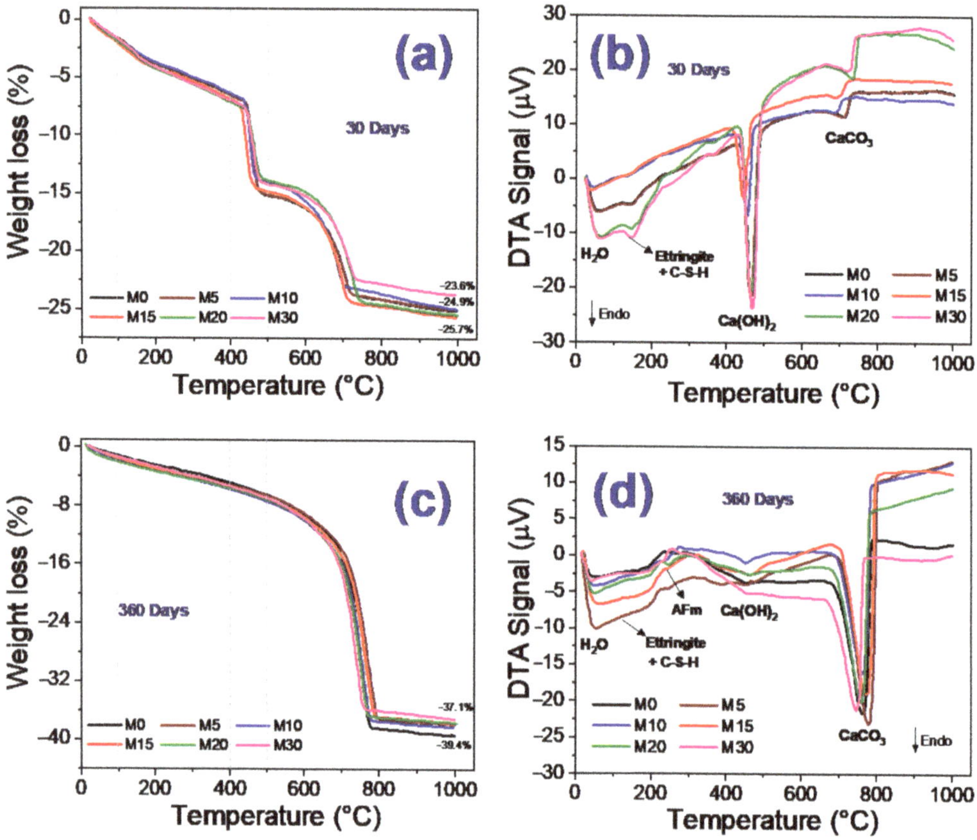

Figure 6. The TGA and DTA curves of mortars subjected to curing periods in an internal environment for 30 days (a,b) and 360 days (c,d). All the samples were heated with a heating rate of 12.5 °C/min^{-1} and up to 1000 °C in a nitrogen gas atmosphere.

An increase in loss was observed for all samples studied after 360 days of aging in the internal environment (Figure 6c,d). This result may be related to the exposure of the mortars to the ambient air, causing carbonation from the reaction with CO_2. In fact, the intensity of the endothermic peaks related to calcite decomposition ($CaCO_3$) increases with time (see Figure 6b,d). Moreover, the intensity of the endothermic peaks, from the dehydroxylation of portlandite ($Ca(OH)_2$) decreased significantly with time. This indicates that portlandite was consumed by the carbonation reaction.

For M0 (mortar without KR) the mass loss in the fourth region (i.e, the zone related to calcite decomposition) was approximately 18.6% and 32.7%, at 30 days and 360 days, respectively. However, for the samples with KR (M5, M10, M15, M20, and M30) the losses in this region were smaller, in the range of 9.9–11.6% and 30.4–31.6%, at 30 and 360 days.

This indicates that the formation of CSH was probably favored in the samples with calcined KR and, as a consequence, the carbonation effect was relatively moderate.

Figure 7a,b show the DTA curves of mortars aged naturally in an external environment for 90 days (Figure 7a) and 512 days (Figure 7b). For all compositions, endothermic peaks were observed at around 156 °C and 215 °C, which are related to ettringite and AFm (hemicarbonate and monocarbonate) phases, respectively [39,43]. It is known that calcined KR, in addition to silicon, has a high content of alumina in its composition (see Table 1). Thus, the presence of hemi and monocarbonate compounds are probably associated with the chemical reaction of calcite ($CaCO_3$) from lime, and the extra alumina present in the system is likely due to the addition of kaolin residue. It is important to emphasize that the formation of hemi and monocarbonate compounds inhibits the formation of monosulfate, so the ettringite remains in the system for a longer period of time [44]. The ettringite, hemi, and monocarbonate phases occupy a greater volume of solid hydrates in the mortar pores compared to monosulfates, thus encouraging the filling up of voids and the attainment of mechanical strength [45].

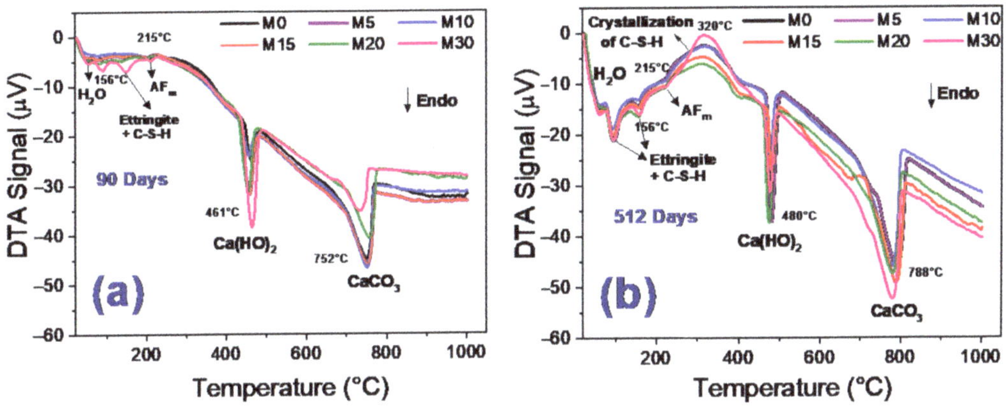

Figure 7. The DTA curves of mortars subjected to curing periods in an external environment for 90 days (**a**) and 512 days (**b**). All the samples were heated with a heating rate of 12.5 °C/min^{-1} in a nitrogen gas atmosphere.

Intense peaks at ~461–480 °C and ~750 °C were also observed, which are associated with portlandite and calcium carbonate. This indicates that all mortars present an excess of free lime. The increase in the curing time from 90 to 512 days caused the displacement and intensification of the peak relative to calcium carbonate. With the displacement, the $CaCO_3$ peak occurred from ~752 °C to ~788 °C. Such behavior is related to the consumption of portlandite by the pozzolanic reaction and carbonation.

In carbonation, CO_2 comes into contact with water and the carbonate elements ($Ca(OH)_2$, C-S-H, C_2S, and C_3S) in the mortar pores. This process depends on relative humidity, temperature, and CO_2 concentration. Furthermore, the inefficient compaction and high water–cement ratio generate high permeability in the material [46]. For mortars cured at 512 days (Figure 7b), it is possible to observe the appearance of exothermic bands between the temperatures of 215–450 °C. These bands refer to the crystallization of CSH.

Figure 8a–f show the CS values and percentages of increase/decrease in the samples submitted to natural aging in the internal (Figure 8a,c,e) and external environments (Figure 8b,d,f). In Figure 8a, it is possible to observe an increase in the compressive strength of samples cured up to 180 days. Such behavior is related to the cement hydration process and the pozzolanic reactions.

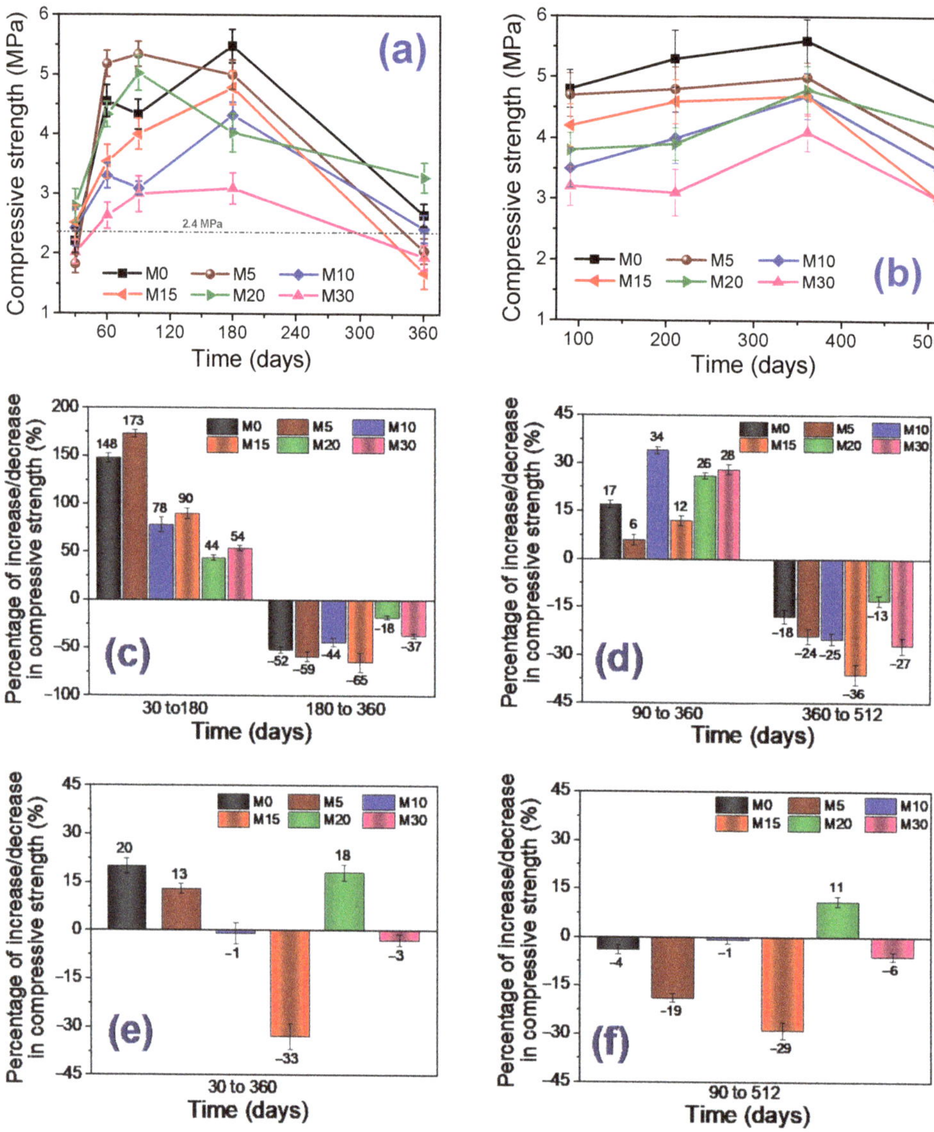

Figure 8. The compressive strength and the percentage of increase/decrease in the aged mortars in the internal environment of the laboratory for 30, 60, 90, 180, and 360 days (**a**,**c**,**e**), and the exposure to the external environment for 90; 210; 360; and 512 days (**b**,**d**,**f**).

The main products formed during the cement hydration process are hydrated calcium silicate (primarily CSH) and portlandite $Ca(OH)_2$. CSH is a non-crystalline product, called agglomerating gel, that provides resistance to mortar and concrete and contributes significantly to the durability of cementitious materials. Moreover, due to the pore refinement process, there is a decrease in the permeability of the system and, therefore, a hinderance to the contact of reactive substances with the cement. On the other hand, portlandite consists of low resistance crystals that are soluble in water and which do not contribute to the material's resistance and durability [6,47,48].

In the pozzolanic reaction, the calcined kaolin residue (metakaolin) reacts with calcium hydroxide (portlandite) to form additional hydrated calcium silicate (secondary CSH), see Equation (2). Figure 9 shows the schematic representation of the microstructure of the material without pozzolan and with the incorporation of pozzolan. The CSH particles are smaller when compared to $Ca(OH)_2$, and have the ability to occupy capillary voids. These particles also have a greater specific area, providing the cement paste with greater mechanical resistance [49].

$$3Ca(OH)_2 + SiO_2 \rightarrow 3CaO \cdot SiO_2 \cdot 3H_2O \tag{2}$$

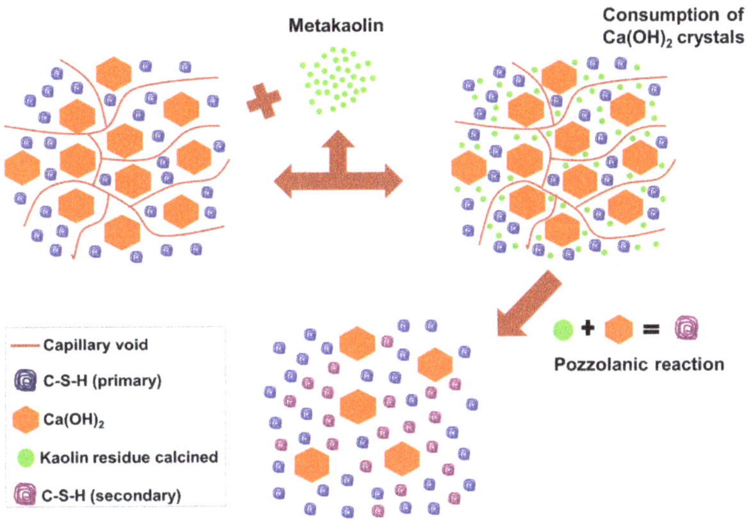

Figure 9. The schematic representation of the microstructure of cementitious material with and without the addition of pozzolan.

Regardless of the aging test type (internal and external environments), M10, M15, and M30 samples showed lower compressive strength values than the reference sample (M0). Moreover, at the end of the aging period in the internal environment (30 to 360 days), there was no resistance gain for M10, M15, and M30 samples. It is noteworthy that the M15 sample showed a 33% reduction in resistance to the initial cure time (30 days), see Figure 8e. These same mortars showed similar behavioral patterns when subjected to aging in an external environment (Figure 8d,f), which indicates that, although it has presented considerable initial gains, the replacement of cement with calcined kaolin residue, in the proportions of 10%, 15%, and 30%, did not lead to an improvement in its durability over time.

In general, there is a considerable decrease in the mechanical strength for all mortars with the evolution of the curing time ranging from 180 to 360 days (internal environment aging: Figure 8a) and from 360 to 512 days (external environment aging: Figure 8b). Such behavior can be associated with the carbonation process. Carbonation occurs due to carbon dioxide gas (CO_2), present in the air, penetrating the mortars through the pores in the material. In the presence of water, CO_2 dissolves and forms CO_3^{2-} ions which react with Ca^{2+} ions to form calcite, as shown in Equation (3). Ca^{2+} ions originate mainly from portlandite.

$$Ca(OH)_2 + CO_2 \overset{H_2O}{\rightarrow} CaCO_3 + H_2O \tag{3}$$

In addition to the portlandite, other hydrated compounds (calcium aluminates and silicates) are also sensitive to CO_2 attack. The reaction between CO_2 and $Ca(OH)_2$ is more favorable from a kinetics viewpoint. For example, the reaction between CO_2 and $Ca(OH)_2$ is three times faster than the reaction between CO_2 and CSH, which is twenty times faster than the reaction between CO_2 and C_2S, and fifty times faster than the reaction between CO_2 and C_3S [17,50]. Figure 10 schematically shows the carbonation process that occurs when CO_2 enters the mortar surface through the pores in the material.

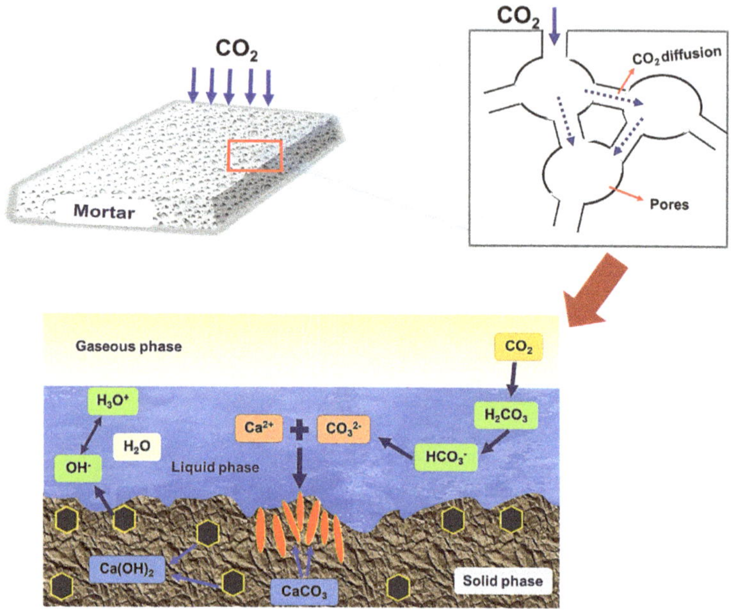

Figure 10. The schematic representation of the carbonation process.

The composition with 20% of residue (M20) presented the lowest percentage of resistance loss among all the compositions, and was the only case that, at the end of the whole process, presented resistance gain in both forms of aging (internal and external environments), see Figure 8e,f. At the end of the 360 days, in the internal environment aging test, the compressive strength of this composition surpassed the MPa of the other compositions (3.3 ± 0.25). This shows that, even with carbonation, the M20 sample degraded less. This behavior is probably associated with the greater packaging of particles, making CO_2 access more difficult and reducing the carbonation effect. Figure 11 shows the water absorption results (WA) of the mortars after the aging test in the internal environment. It is possible to observe that, at the end of the 360 days, M20 presented the lowest value of WA (12.6 ± 0.3)% compared with the other compositions. In this way, it is reasonable to infer that the M20 sample presented minor porosity, at least between 90 and 360 days of acting under the internal environment. Such a conclusion strengthens the idea of CO_2 access reducing the carbonation effect.

Therefore, the porosity present in the mortars and concretes is one factor that can affect durability and stability since it facilitates the aggressive agent diffusibility to the material, such as carbon dioxide, acidic solutions, and sulfides. Thus, it is essential to highlight the importance of studying the degradation behavior of mortars containing residues, since the result of their resistance, which is generally measured after curing in 28 days, does not always indicate the real behavior of this material over time. Furthermore, it is essential to consider the content of substitution for the use of residues in mortars. The M0, M10, and M20 samples presented resistance values within the minimum limit (2.4 MPa) established by ASTM C 270 [51] in both types of aging tests (internal and external environments).

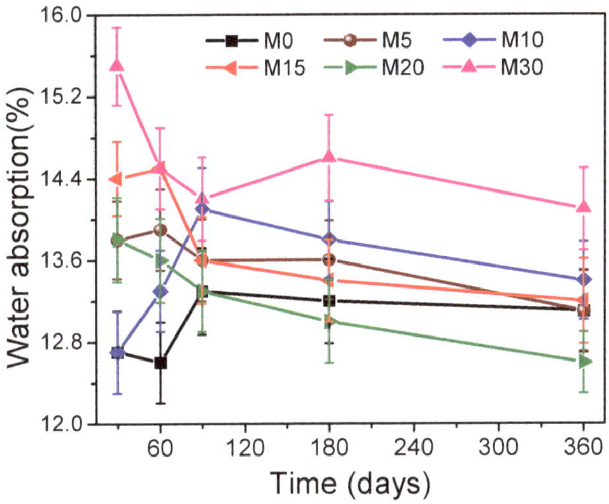

Figure 11. The water absorption of the aged mortars in the internal environment of the laboratory for 30, 60, 90, 180, and 360 days.

4. Conclusions

Based on our results, it can be concluded that the partial replacement of Portland cement with kaolin residue in coating mortars is feasible. The addition of calcined kaolin residue in the mortars reduced the mechanical resistance to compression in practically all substitution levels. However, mortars with a 10% and 20% substitution presented resistance values within the minimum limit of 2.4 MPa established by ASTM C 270. The content of replacing cement with a kaolin residue influenced the durability behavior of the mortars. Incorporating the residue in the substitution proportions of 5%, 10%, 15%, and 30% did not improve their mechanical resistance or durability. The composition with 20% of kaolin residues presented, again, compressive strength at the end of the entire aging process to the initial evaluation stages. And at the end of 360 days of aging in an internal environment, the compressive strength of this composition surpassed the MPa of the other compositions (3.3 ± 0.25).

Author Contributions: Methodology, software, data curation, and formal analysis, F.P.d.C., S.L.D.B. and J.V.F.; conceptualization, supervision, funding acquisition, and project administration, G.d.A.N. and R.R.M.; formal analysis and writing—review and editing, A.M.R. All authors have read and agreed to the published version of the manuscript.

Funding: This research was funded by Coordenação de Aperfeiçoamento de Pessoal de Nível Superior (CAPES), grant number 88887.597478/2021-00, and by Conselho Nacional de Desenvolvimento Científico e Tecnológico (CNPq), grant number 140211/2021-7.

Data Availability Statement: Not applicable.

Acknowledgments: The authors would like to thank the Laboratório de Tecnologia do Materiais (LTM-UAEMa/UFCG) for the infrastructure to carry out tests.

Conflicts of Interest: The authors declare no conflict of interest.

References

1. Amaral, L.F.; Girondi Delaqua, G.C.; Nicolite, M.; Marvila, M.T.; de Azevedo, A.R.G.; Alexandre, J.; Fontes Vieira, C.M.; Monteiro, S.N. Eco-friendly mortars with addition of ornamental stone waste—A mathematical model approach for granulometric optimization. *J. Clean. Prod.* **2020**, *248*, 119283. [CrossRef]
2. Cree, D.; Pliya, P. Effect of elevated temperature on eggshell, eggshell powder and eggshell powder mortars for masonry applications. *J. Build. Eng.* **2019**, *26*, 100852. [CrossRef]

3. McLellan, B.C.; Williams, R.P.; Lay, J.; Van Riessen, A.; Corder, G.D. Costs and carbon emissions for geopolymer pastes in comparison to ordinary portland cement. *J. Clean. Prod.* **2011**, *19*, 1080–1090. [CrossRef]
4. Chen, Z.; Poon, C.S. Comparative studies on the effects of sewage sludge ash and fly ash on cement hydration and properties of cement mortars. *Constr. Build. Mater.* **2017**, *154*, 791–803. [CrossRef]
5. Bilir, T.; Gencel, O.; Topcu, I. Properties of mortars with fly ash as fine aggregate. *Constr. Build. Mater.* **2015**, *93*, 782–789. [CrossRef]
6. Hsu, S.; Chi, M.; Huang, R. Effect of fineness and replacement ratio of ground fly ash on properties of blended cement mortar. *Constr. Build. Mater.* **2018**, *176*, 250–258. [CrossRef]
7. Onn, C.C.; Mo, K.H.; Radwan, M.K.; Liew, W.H.; Ng, C.G.; Yusoff, S. Strength, carbon footprint and cost considerations of mortar blends with high volume ground granulated blast furnace slag. *Sustainability* **2019**, *11*, 7194. [CrossRef]
8. Seifi, S.; Sebaibi, N.; Levacher, D.; Boutouil, M. Mechanical performance of a dry mortar without cement, based on paper fly ash and blast furnace slag. *J. Build. Eng.* **2019**, *22*, 113–121. [CrossRef]
9. Rakhimova, N.; Rakhimov, R.Z. Alkali-activated cements and mortars based on blast furnace slag and red clay brick waste. *Mater. Des.* **2015**, *85*, 324–331. [CrossRef]
10. Pliya, P.; Cree, D. Limestone derived eggshell powder as a replacement in Portland cement mortar. *Constr. Build. Mater.* **2015**, *95*, 1–9. [CrossRef]
11. Ahmad, M.; Chen, B. Effect of silica fume and basalt fiber on the mechanical properties and microstructure of magnesium phosphate cement (MPC) mortar. *Constr. Build. Mater.* **2018**, *190*, 466–478. [CrossRef]
12. Benli, A. Mechanical and durability properties of self-compacting mortars containing binary and ternary mixes of fly ash and silica fume. *Struct. Concr.* **2019**, *20*, 1096–1108. [CrossRef]
13. Gupta, S.; Kua, H.W. Combination of Biochar and Silica Fume as Partial Cement Replacement in Mortar: Performance Evaluation Under Normal and Elevated Temperature. *Waste Biomass Valorization* **2020**, *11*, 2807–2824. [CrossRef]
14. Chouhan, H.S.; Kalla, P.; Nagar, R.; Gautam, P.K. Gainful utilization of dimensional limestone waste as fine aggregate in cement mortar mixes. *Constr. Build. Mater.* **2019**, *221*, 363–374. [CrossRef]
15. Pozo-Antonio, J. Evolution of mechanical properties and drying shrinkage in lime-based and lime cement-based mortars with pure limestone aggregate. *Constr. Build. Mater.* **2015**, *77*, 472–478. [CrossRef]
16. Hanjitsuwan, S.; Phoo-ngernkham, T.; Li, L.Y.; Damrongwiriyanupap, N.; Chindaprasirt, P. Strength development and durability of alkali-activated fly ash mortar with calcium carbide residue as additive. *Constr. Build. Mater.* **2018**, *162*, 714–723. [CrossRef]
17. Sabir, B.; Wild, S.; Bai, J. Metakaolin and calcined clays as pozzolans for concrete: A review. *Cem. Concr. Compos.* **2001**, *23*, 441–454. [CrossRef]
18. Koutník, P.; Soukup, A.; Bezucha, P.; Šafář, J.; Kohout, J. Low viscosity metakaolinite based geopolymer binders. *Constr. Build. Mater.* **2020**, *230*, 116978. [CrossRef]
19. Silva, A.S.; Gameiro, A.; Grilo, J.; Veiga, R.; Velosa, A. Long-term behavior of lime-metakaolin pastes at ambient temperature and humid curing condition. *Appl. Clay Sci.* **2014**, *88–89*, 49–55. [CrossRef]
20. Al-Akhras, N.M. Durability of metakaolin concrete to sulfate attack. *Cem. Concr. Res.* **2006**, *36*, 1727–1734. [CrossRef]
21. Güneyisi, E.; Gesoğlu, M.; Mermerdaş, K. Improving strength, drying shrinkage, and pore structure of concrete using metakaolin. *Mater. Struct.* **2008**, *41*, 937–949. [CrossRef]
22. Bassuoni, M.T.; Nehdi, M.L. Resistance of self-consolidating concrete to sulfuric acid attack with consecutive pH reduction. *Cem. Concr. Res.* **2007**, *37*, 1070–1084. [CrossRef]
23. Goyal, S.; Kumar, M.; Sidhu, D.S.; Bhattacharjee, B. Resistance of Mineral Admixture Concrete to Acid Attack. *J. Adv. Concr. Technol.* **2009**, *7*, 273–283. [CrossRef]
24. Senhadji, Y.; Escadeillas, G.; Khelafi, H.; Mouli, M.; Benosman, A.S. Evaluation of natural pozzolan for use as supplementary cementitious material. *Eur. J. Environ. Civ. Eng.* **2012**, *16*, 77–96. [CrossRef]
25. ASTM International. *ASTM C150/C150M-20, Standard Specification for Portland Cement*; ASTM International: West Conshohocken, PA, USA, 2020.
26. ASTM International. *ASTM C1437, Standard Test Method for Flow of Hydraulic Cement Mortar*; ASTM International: West Conshohocken, PA, USA, 2020.
27. ASTM International. *ASTM C311-18-Standard Test Methods for Sampling and Testing Fly Ash or Natural Pozzolans for Use in Portland-Cement Concrete*; ASTM International: West Conshohocken, PA, USA, 2018.
28. ASTM International. *ASTM C39/C39M-20-Standard Test Method for Compressive Strength of Cylindrical Concrete Specimens*; ASTM International: West Conshohocken, PA, USA, 2020.
29. Huang, Y.; Deng, J.; Wang, W.; Feng, Q.; Xu, Z. Preliminary investigation of pozzolanic properties of calcined waste kaolin. *Medziagotyra* **2018**, *24*, 177–184. [CrossRef]
30. Almeida, E.P.; Carreiro, M.E.A.; Rodrigues, A.M.; Ferreira, H.S.; Santana, L.N.L.; Menezes, R.R.; Neves, G.A. A new eco-friendly mass formulation based on industrial mining residues for the manufacture of ceramic tiles. *Ceram. Int.* **2020**. [CrossRef]
31. Da Silva, V.J.; da Silva, M.F.; Gonçalves, W.P.; de Menezes, R.R.; de Araújo Neves, G.; de Lucena Lira, H.; de Lima Santana, L.N. Porous mullite blocks with compositions containing kaolin and alumina waste. *Ceram. Int.* **2016**, *42*, 15471–15478. [CrossRef]
32. Bui, D.D.; Hu, J.; Stroeven, P. Particle size effect on the strength of rice husk ash blended gap-graded Portland cement concrete. *Cem. Concr. Compos.* **2005**, *27*, 357–366. [CrossRef]

33. Kiattikomol, K.; Jaturapitakkul, C.; Songpiriyakij, S.; Chutubtim, S. A study of ground coarse fly ashes with different finenesses from various sources as pozzolanic materials. *Cem. Concr. Compos.* **2001**, *23*, 335–343. [CrossRef]
34. Da Costa, F.P.; da Silva Morais, C.R.; Rodrigues, A.M. Sustainable glass-ceramic foams manufactured from waste glass bottles and bentonite. *Ceram. Int.* **2020**, *46*, 17957–17961. [CrossRef]
35. Da Costa, F.P.; da Silva Morais, C.R.; Pinto, H.C.; Rodrigues, A.M. Microstructure and physico-mechanical properties of Al2O3-doped sustainable glass-ceramic foams. *Mater. Chem. Phys.* **2020**, *256*, 123612. [CrossRef]
36. ASTM International. *ASTM C618-19-Standard Specification for Coal Fly Ash and Raw or Calcined Natural Pozzolan for Use in Concrete 1*; ASTM International: West Conshohocken, PA, USA, 2019. [CrossRef]
37. Sun, C.; Sun, M.; Tao, T.; Qu, F.; Wang, G.; Zhang, P.; Li, Y.; Duan, J. Chloride binding capacity and its effect on the microstructure of mortar made with marine sand. *Sustainability* **2021**, *13*, 4169. [CrossRef]
38. Michel, M.; Georgin, J.F.; Ambroise, J. Improving the mechanical performance of high-grade slag cement by the addition of Portland cement and sulfoaluminate cement. *Constr. Build. Mater.* **2012**, *37*, 291–300. [CrossRef]
39. De Weerdt, K.; Kjellsen, K.O.; Sellevold, E.; Justnes, H. Synergy between fly ash and limestone powder in ternary cements. *Cem. Concr. Compos.* **2011**, *33*, 30–38. [CrossRef]
40. Song, H.; Jeong, Y.; Bae, S.; Jun, Y.; Yoon, S.; Oh, J.E. A study of thermal decomposition of phases in cementitious systems using HT-XRD and TG. *Constr. Build. Mater.* **2018**, *169*, 648–661. [CrossRef]
41. Liu, C.; Yang, L.; Wang, F.; Hu, S. Enhace the durability of heat-cured mortars by internal curing and pozzolanic activity of lightweight fine aggregates. *Constr. Build. Mater.* **2021**, *270*, 121439. [CrossRef]
42. Rupasinghe, M.; Nicolas, R.S.; Mendis, P.; Sofi, M.; Ngo, T. Investigation of strength and hydration characteristics in nano-silica incorporated cement paste. *Cem. Concr. Compos.* **2017**, *80*, 17–30. [CrossRef]
43. Krishnan, S.; Bishnoi, S. Understanding the hydration of dolomite in cementitious systems with reactive aluminosilicates such as calcined clay. *Cem. Concr. Res.* **2018**, *108*, 116–128. [CrossRef]
44. Antoni, M.; Rossen, J.; Martirena, F.; Scrivener, K. Cement substitution by a combination of metakaolin and limestone. *Cem. Concr. Res.* **2012**, *42*, 1579–1589. [CrossRef]
45. Zajac, M.; Rossberg, A.; Le Saout, G.; Lothenbach, B. Influence of limestone and anhydrite on the hydration of Portland cements. *Cem. Concr. Compos.* **2014**, *46*, 99–108. [CrossRef]
46. Nasser, A.; Clément, A.; Laurens, S.; Castel, A. Influence of steel–concrete interface condition on galvanic corrosion currents in carbonated concrete. *Corros. Sci.* **2010**, *52*, 2878–2890. [CrossRef]
47. Heukamp, F.H.; Ulm, F.-J.; Germaine, J.T. Poroplastic properties of calcium-leached cement-based materials. *Cem. Concr. Res.* **2003**, *33*, 1155–1173. [CrossRef]
48. Singh, M.; Garg, M. Reactive pozzolana from Indian clays—their use in cement mortars. *Cem. Concr. Res.* **2006**, *36*, 1903–1907. [CrossRef]
49. Morandeau, A.; Thiery, M.; Dangla, P. Investigation of the carbonation mechanism of CH and CSH in terms of kinetics, microstructure changes and moisture properties. *Cem. Concr. Res.* **2014**, *56*, 153–170. [CrossRef]
50. Aguirre-Guerrero, A.M.; Mejía-de-Gutiérrez, R.; Montês-Correia, M.J.R. Corrosion performance of blended concretes exposed to different aggressive environments. *Constr. Build. Mater.* **2016**, *121*, 704–716. [CrossRef]
51. ASTM International. *ASTM C270-19-Standard Specification for Mortar for Unit Masonry*; ASTM International: West Conshohocken, PA, USA, 2019.

MDPI
St. Alban-Anlage 66
4052 Basel
Switzerland
Tel. +41 61 683 77 34
Fax +41 61 302 89 18
www.mdpi.com

Sustainability Editorial Office
E-mail: sustainability@mdpi.com
www.mdpi.com/journal/sustainability